Ethics in Medicine

Shabih H. Zaidi

Ethics in Medicine

 Springer

Shabih H. Zaidi
American University of Barbados
Birmingham
UK

ISBN 978-3-319-01043-4   ISBN 978-3-319-01044-1   (eBook)
DOI 10.1007/978-3-319-01044-1
Springer Cham Heidelberg New York Dordrecht London

Library of Congress Control Number: 2013956433

Printed on acid-free paper

Springer is part of Springer Science+Business Media (www.springer.com)

*Humbly dedicated to
Mohammad- o- Aley Mohammad
the ultimate teachers of ethics*

*With thanks to my parents
who brought me up to follow them*

Foreword

Medical science has been around for a long time; in fact, it has probably existed since the first man became sick. Over time it has evolved and become specialised. The medical profession is one of the most respectable ones in the world. Due to the nature of its work, curing bodies and saving lives, humanity has always felt close to this livelihood.

Ethics is a branch of philosophy. Ethical influence has changed the way human beings deal with many issues. It continues to hold its ground in the modern world. Ethics is a means of monitoring man's activities. It strives to keep man aligned with divine law as *'corruption has appeared in land and sea because of the doings of people's hands'* (Quran; 30:40). There can be two insights to the word 'Ethics': ethics as morality or the human conduct, and ethics as philosophy of good actions. The former deals with what human beings should do and the latter deals with why they should do what they should do. In other words, one is the law of good life and the other is the reasoning behind that law.

Ethical arguments exist in all social sciences. However, its position in medicine is undeniable. The weight of ethics in medicine increased in the twentieth century and persists to do so in the new millennium. Medical Ethics is a growing topic as new discoveries are constantly being made in the field. Furthermore, we are living in an all changing world in terms of ideologies and awareness. Religion is playing a more important role in Medical Ethics than ever before. The responsibility of all researchers and writers on this subject matter is crucial; for their contribution will be a torch for future generations.

Dr. Shabih Haidar Zaidi covers a wide range of topics in this book, *Medical Ethics*. He gives the view of leading medical figures, philosophers and religious experts. The reader will appreciate the extensive research the author has carried out in compiling such a work. It makes fine comparisons and offers thorough analyses.

My acquaintance with Dr. Shabih Haidar Zaidi is not recent. Ever since I have known him, I have admired his intellect, professionalism, integrity and academic credibility. He has served as an expert in many different medical establishments. He continues to progress at the peak of his professional career. As a student for life one should not stop learning and improving. Dr. Zaidi is a perfect example of this constant pursuit of knowledge. In this book, Dr. Zaidi has explored the contemporary issues pertaining Medical Ethics. I wish him the best for his new position as the Dean of the Medical School in Barbados.

London, England

(HujatulIslam Molana)
Ali Raza Rizvi
President Shia Ulema , Europe

Contents

Introduction

Medical ethics and medical profession are just as inseparable from each other as motherhood and mother. Formal teaching of medical ethics is a relatively new phenomenon. Previously innumerable generations of physicians imbibed ethical values from their masters and mentors through long associations, apprenticeship, observation and a process of osmosis.

Besides, the life was simple as being a Good Samaritan and possessing simple bed side manners, like greeting the patient with formal salutation, treating him with empathy, professional efficiency were more than adequate.

Since the decline and fall of the 'paternalistic approach', and introduction of Managed Health Care, as indeed the explosion of knowledge and technology, the whole concept of medical ethics has undergone a sea change. Now, the physician is a 'health provider' and the patient a 'client'. The profession has become a trade. A contract has to be drawn between a client and the server each time a service is provided. Caution and precaution are the natural ingredients of such a relationship, resulting in defensive medicine, practiced these days.

What is Ethics? Well, just like logic, metaphysics, politics and aesthetics, ethics is a branch of philosophy. Medical ethics is the application of normative principles of ethics in medical sciences. Morality and ethics are interchangeable terms, albeit ethics is primarily the name given to the theoretical principles as compared with morality, which is an applied form of those principles. There are two moral pathways to follow: Religious or Secular. Both have the same basic agenda namely the welfare of mankind. Just as it is not essential that every religious person must hold the high ground purely on the basis of faith, it is also not mandatory that a secular person may not enjoy the same status for the lack of his faith. Both have equal rights to hold the moral high ground based upon their deeds rather than their beliefs.

The first recorded code of conduct was the Hammurabi code written in 1792 BC. It was followed by the ageless Hippocratic doctrine (460 BC–390 BC). Many of us have taken that as a professional oath at our graduation. Further developments in this field have happened since those ancient times. Faith based oaths are now in common use in many nations, particularly in the Islamic world.

Mankind has gradually progressed and marched forward over several millennia. The first millennium was the age of Philosophers, the second of Religions, and the present millennium is the age of information and technological advancement. Each had its own codes of conduct; none more lasting than those of the second.

Applied ethics has continued to evolve; but the fundamental principles have not changed. In other words, universal norms and virtues like, truth, honesty, justice, loyalty, integrity, generosity, valour and courage etc. have stood the test of time; however their importance may change with the era. For instance in the ancient times, individual bravery, art of warfare, physical strength etc. preceded other values. Today, individual valour and art of one-to-one combat is replaced by intelligence, collective wisdom or the creation, development and the use of modern technology etc.

Ancient civilisations like the Incas, Aztecs, Mayas the Indo-Harrapans

Babylonians, Mesopotamians, Assyrian, the Chinese, and the Egyptians had their own codes of conduct. They served their purpose. A challenge to the established doctrines was met with fierce opposition, oppression even annihilation of the reformer. But the battle between the good and evil went on. Sometimes the saints, savants, and the Prophets had to take drastic actions. Moses fought against the worst tyranny of the Pharoes and gave the world, divine Ten Commandments.

Then came the Greeks, first as the Sophists like Pythagoras and Thales, then all time greats namely Socrates, Plato, and Aristotle. They all strived hard to advance the cause of good over evil. It was Aristotle though, who has left indelible marks in history of ethics through his book called Nichmacean ethics.

Judeo-Christian ethics prevails across the globe following the code of conduct given by Moses, and later on by Jesus Christ. There are more Christians in the world than any other monotheistic faith followers. The message of Jesus was universal and continues to serve the world. Judaism, on the other hand remains static, as neither does it allow conversion nor does it believe in preaching. One is born as a Jew, but cannot acquire the Jewish faith by choice.

Is it not amazing that the very cradle of civilisation is located by the river Euphrates and all three Abrahamic monotheistic religions dawned between the Red Sea and the Dead Sea!

Islam came through Mohammad in 610 AD and dominated the world, particularly Europe, until 1492 when Ferdinand and Isabella of Aragon threw out the last Muslim rulers from Andalusia, also resulting in Jewish diaspora to Europe. Arab philosopher Ibne Rushd introduced Aristotle to Thomas Acquinos and Maimonides, who forwarded it to Europe of the Middle Ages, thus bridging the gap between the Greek period and Renaissance. Christopher Columbus discovered America the same fateful year. Revival of Islam since the Iranian revolution brought in by Ayatullah Khomeni is a huge phenomenon, which has changed many a thing in the contemporary world. 1979 may be labelled as watershed year by the historians to come. There are more than 1.2 Billion Muslims in the world today who follow the ethical and moral guidance taught by Quran and Sunnah.

For generations, the church had authoritatively controlled the human thought, its propagation and documentation. So, with the Renaissance the philosophers became independent, rebellious, even anti-church. Therefore, many thinkers openly criticised the religious mandate on moral principles and introduced their own. From Bacon to Kant and Voltaire to Russell, a multitude of formidable philosophers have transformed the human thought. Machiavelli and Netsche had different ideas. In the field of Medical ethics, Aristotle, Immanuel Kant, Bentham and Mills continue to dominate. Modern philosophers like Mcintyre, Singer and Dawkins are names to reckon with. Some are pro-life, and believers of a faith, others anti-God. And yet there are some who call Higgs Boson discovered in July 2012, as the God particle!

In 1990s, Childress and Beauchamp, two renowned American ethicists formally introduced the fundamental pillars of medical ethics. They are Autonomy, Beneficence, Non- malificence and Justice. The spine of autonomy is: 'Informed consent'. Besides, human dignity, respect for human rights and informed decision making are the fundamental rights of a patient, duly balanced with obligations. Justice remains the most fundamental block of medical ethics. It functions like a string in a rosary, holding all the beads together.

A myriad of ethical dilemmas have cropped up in recent years. No physician can be aware of them all, or indeed know their answers. Many situations are by products of technology such as cloning and the Human Genome Project. One has the potential to become an evil act, the other a shining illustration of a good action. While human cloning is to be condemned and is actually forbidden, the later is duly encouraged, indeed rapidly being promoted across the cultures, and faiths. Employment of human embryos in harvesting stem cells was an issue until recently, but new technologies have nearly solved that dilemma. Stem cells and Eugenics may change the whole panorama of pathology in the future years.

There are many slippery slopes in the world of morality and ethics. Some differences between a good and an evil act are paper thin!

Organ transplantation is now a norm, but many societies are riddled with the problem of organ theft and sale. The major question today is that if a token payment may indeed be allowed to the donor as a token of thanks rather than the price of an organ. Presumed consent for organ donation with the provision of opting out is another contemporary issue requiring further debate. Some countries have legislated cadaveric donation, which is commendable, but raising fears of unduly prolonging life for organ harvesting. Death is inevitable, and once the Brain and Somatic death have been confirmed, the dyeing process may not be delayed. The role of an ethics committee cannot be overemphasised in such situations.

The field of reproductive health is surrounded with controversies. For instance, abortion. Some faiths outright condemn it, others have put strictures and caveats, on its application. Highly innovative technologies have enabled the scientists to establish a major foetal disability in very early stages of pregnancy through a process called Pre-genetic diagnosis. If positive, should the mother have an option to abort, and if so at what stage. If not, then who, besides the mother should bear the economic cost of the handicapped child, who may live for many long years. Will such an action compromise the provision of essential services to many others who may otherwise live a normal, productive healthy life. Services in the light of limited resources is a major ethical debate covering subjects like calculation of DALYs and QALYs.

Infertility can be devastating for families in certain patriarchal societies. Adoption is one way, but new technologies of Third Party Gamete transfer through a temporary marriage is currently under investigation in some Islamic countries. It allows the parents to have their own biological off spring.

Sperm banks have been set up in many countries, particularly in the USA, which exports a huge sperm load to many countries. Can it lead to incest? Avoiding incest is a major issue in some cultures. Jewish centres insist on knowing the identity of the donor, and prefer a non Jewish donor to avoid such a mishap.

Then there is the question of euthanasia, assisted dyeing, the controversial palliative care pathways, and the final visit to some dedicated places for a planned death. Many people have reservations on these issues.

The lessons learnt from the notorious Dr. Mangle and his reincarnation in modern times in the shape and form of Dr. Shipman are never to be forgotten.

Assisted life prolongation is another major issue, particularly in those countries where health care is an individual's responsibility and the burden of disease and possible disability may not be shared by the state.

Clinical trials of new molecules developed in the labs on human volunteers is essential; but there are many unethical practices in vogue when phase 4 trials are conducted. Placebos have been employed in the controls despite the proven efficacy of a drug, thereby raising the very basic question of abuse of authority and total neglect of the principles of Good Medical Practice. Distributive justice demands that such drugs should be made freely available to the subjects on whom they are tried, but many times they were withdrawn or made unaffordable. In many countries, collaborative research programs are involved in the abuse of children as well as adults, without their informed consent, or informed decision making. That must be stopped.

Many disturbing reports keep surfacing in the media about the pathetic situation in foster care homes for children, and the care homes for the elderly, the sick and the disabled. The nursing profession is often blamed for such neglects, as indeed the inpatient care in some hospitals. Mid Staffordshire hospital tragedy is one blatant example of recent times.

Many countries are involved in employing the prisoners of conscience and others in their custody not just for human experimentation, but if they are condemned, sometimes due to nefarious reasons, their organs are hastily removed and sold off to the rich waiting recipients. Doctors are sometimes coerced to be a part of torturing inmates in some countries, which is not just unethical, but outright condemnable heinous crime on the part of the medical personnel involved. Medical profession must resist and stop such malpractices.

Fundamental principles of medical care, medical ethics and dilemmas faced by the profession are universal but all faiths and cultures have their own values, beliefs, and traditions that play an important role in the final outcome. Therefore some philosophers believe that Moral Relativism should be taken into account to explain such practices as surrogate consent instead of an informed personal consent, in certain societies. Cultural anthropology is an essential component of modern medical ethics. It's awareness is of great benefit to a medical personnel, as this profession is sometimes aptly called a profession without geographical borders.

All three monotheistic religions have taught the same principles through their gospels and preachings, namely 'Bid good, Forbid evil'. Many dilemmas are discussed in the light of religious mandates, and extrapolated to meet the present times.

Investigative medical writing can be controversial, but innovative, though provoking, inspirational, emotive and educational. All faiths and philosophies must be duly respected. They all mean to reform the society. And the society is a collection of people living in certain communes in a certain order, following certain rules and regulations.

There is a triangle of ethics which comprises of knowledge, skills and conscience. Every one knows that Knowledge is power, skills enable one to apply the knowledge for a practical purpose and conscience determines whether the combination of knowledge and skills be employed for the good or an evil purpose. Since conscience is an invisible entity, its tangible form is justice. Therefore if an act is just it would be ethical if unjust it has to be evil. The radiation therapy cures cancer but can also demolish mankind. The technology was developed for a good cause, but its application in the development of a nuclear bomb is obviously evil.

These essays were written over several years while working in the university affiliated NHS hospitals in UK. Besides, many years of practicing medicine in different cultures and countries have enriched the experience a great deal, as did the passion for reading theology, philosophy, literature, poetry, history and anthropology. They are all duly reflected in varying measures in these writings. There are numerous text books, monographs, conference proceedings, even legislative codes available in the libraries and on the net. I have neither the intention nor the competence to compete with those authoritative works. The best way to learn is to teach. Some of the essays are indeed my lectures given in different academic institutions in different countries; hence the reader my notice repetition of some fundamental facts. It is usually necessary to bring the past in direct continuity with the present subject, hence the repetition may be overlooked. As a student of medical ethics one is always keen to learn and share the personal experiences with colleagues as well as general readers. And that is my motive behind these essays. They are written in a simple narrative style for leisurely reading, as indeed for a student of medical ethics. To many, philosophy can be somewhat daunting and overbearing, but informal essays like these may defy that notion.

During my research, over several years, I have benefitted with many books, monographs, articles, papers as well as speeches from scholars. I have mentioned most of the references wherever applicable. The noble Quran, Prophets' traditions, Nahaj al- Balagah and Saheefa e Kamila have been my main sources of Islamic thought, principles and practices. I have duly benefitted with both Old and the New testament in parts, and many a scholarly works of masters from Plato and Aristotle to Russell and AnnemarrieShmimell to Beacuchamp, Childress, Singer, Dawkins and McIntyre, whose debt upon me is ever lasting. I have richly learnt from SahiBukhari, Ibne MaaJa, Trimizi, Mishakat shareef, Usool e Kaafi, Bahar ul Anwar and writings or lectures of such greats as Syed Maududi, Ayatullal Ul uzma Imam Khomeini, Agha Baqar Al Sadr, Ayatullah ul uzma Agha Syed Sistani, Ayatullah ul uzma Agha Basheer Najafi, Agha Baheshti, Ayatullah uzma Agha Ali Khamenie, AARizvi, Mojen Momen, Rasheed Turabi, Kalbe Sadiq, Talib Jauhari, Syed Moussavi, Aqeel Ghravi, Ai Raza Rizvi, Bagher Larejani, Fazellah Zahedi, Mansureh Senaei, Marcia Inhorn, Zaki Hasan, Maqbool Jafri, Shabeehul Hasan, just to name a few scholars, and ulema. In fact the list is much too exhaustive, but I thank them all whole heartedly. Shaukat Jawaid of *Pakistan Journalists Association* and Mr. Jamsheyd Mirza of Royal Book Co. are dear old friends. I am grateful to them both for allowing me to review and include here some of my own articles and a few other writers from our book called Medical ethics in contemporary era. Some things have radically changed since 1995 and were in dire need of updating.

Finally let us recall that, informal education commences at birth and ends with death. My journey continues....

My last note of gratitude is to Springer, who have graciously accepted to publish my meagre efforts in the form of this book. I find no words adequate enough to thank Sylvana Freyberg in particular, who has obliged me by editing my second book this year. I remain in her perpetual debt.

My particular thanks are due to the Imamia Medics International for honouring me with the editorship of their Ethics newsletter, and allowing me to use some of my publications in this book.

West Midlands, UK   Shabih Zaidi

Ethics Through the Ages

1

1.1 Early Period

Medical ethics is an applied form of ethics, which is a branch of philosophy. hiloso-phy is a study of thoughts, principles, theories, actions, and a way of discussing, explaining, defining, solving a myriad of dilemmas through individual or collective wisdom. It is also a source of lasting mental joy.

'There is a pleasure in philosophy', wrote the renowned historian Will Durant, 'and a lure even in the mirages of metaphysics, which every student faces until the coarse necessities of physical existence drag him from the heights of thought into the depth of economic strife and gain.' Indeed, it is so true that worldly pursuits do not permit one to indulge into the luxury of thought and philosophy. One is so exhausted chasing their livelihood that neither time nor energy are left for one to relax and reflect upon the delicate and intricate patterns and prismatic colours of metaphysics.

Perhaps that is the reason that before the induction of the machine age, a number of philosophers and thinkers emerged and left their indelible impressions on human history. Mechanisation, while improving the economic lot, destroyed such virtues of human beings as 'thought', and 'reason'.

Thinking is the ultimate mental exercise. When one ventures into the thinking process, one goes into gradually deepening layers – from reminiscences of the past to the joys or sorrows of the present and dreams of the future. It is, however, under debate whether any element like 'the present' does actually exist. These thinkers believe that the moment you have taken a breath, or a thought has crossed your mind, you have already moved ahead into the future. They do not give importance to the present, only to the past, which is known and familiar, and the future, which is unknown and unfamiliar but full of hopes and dreams – but that is another meta-physical debate, best left for other times.

All philosophers and all thought-provoking exercises demand a code of conduct – that is 'ethics'. Indeed, philosophy includes five fields of study and discourse: eth-ics, logic, aesthetics, politics, and metaphysics. I believe theology should be a part of this package, but many argue that theology and ethics are the same, which is

S.H. Zaidi, *Ethics in Medicine*,
DOI 10.1007/978-3-319-01044-1_1, © Springer International Publishing Switzerland 2014

debatable because religion does not guarantee morality, nor does lack of faith certify immorality.

Durant defines 'logic' as the study of the ideal method of thought and research, observation and introspection, deduction and induction, hypothesis and experiment, analysis and synthesis. 'Ethics is the study of ideal form or beauty.' It is the philosophy of art. It is the study of ideal conduct. 'The highest knowledge', according to the greatest of all philosophers, Socrates, is, 'the knowledge of the good and the evil; the knowledge of wisdom of life.'

The word *ethic* is derived from the Greek word *ethos*. It is an ancient branch of humanities that deals with human attitudes towards many issues related to norms, behaviour, attitudes, thinking, actions, and a myriad of other day-to-day as well as long-term matters. It is a study of human attitude, reactions, behaviour, principles, rules and norms. It is a science as well as art. Just like the human mind, it has evolved over several millennia.

The study of philosophy teaches us the norms of life, varying from basic human instincts to complex stories of human attitude, which many look like a maze of mysteries, often shrouded in uncertainties. It guides us through the dark alleys of life, as it indeed prepares us to meet the ultimate truth, that is, death.

Philosophy has many prismatic colours. It changes its patterns akin to a kaleidoscope when seen in the light of outer influences, for example, customs and culture. Its understanding is directly related to the level of human intellect and its interpretation could be simple and straightforward or intriguing and beguiling, depending on the matter at hand. It never ceases to amaze its student, and its message is endearing as well as enduring.

One branch of philosophy is 'moral philosophy.' It involves analysis and synthesis of the principles of 'good' and 'evil', 'right' and 'wrong.' The perpetual conflict between '*khair 0-Shar*' is duly studied in moral philosophy.

Although there are subtle differences between morality and ethics, they both deal with human thought on the question of good and evil. Ethics deals with the theories and underlying principles of values and norms and their basic perception and justification.

Moral philosophy deals with customs, traditions and normative behaviour of individuals or groups of people in a given society. Both these studies provide the evil and us with the knowledge of dealing with issues on the basis of differentiating them from the good; hence, often both of these terms are used interchangeably.

Two moral theories have flourished over millennia. The more ancient of the two states that morality is a virtue that is built into the character of a person but is duly influenced by nurturing. In other words, it is affected deeply by prevailing circumstances, the status of the society and socioeconomic matters by the grant and upbringing. It is also affected by the education, training and, most important, the culture of the society one is born into and lives in.

The other more acceptable doctrine of morality states that although one may be virtuous and carry all good values that make one ethical and moral, a religion is necessary and should draw the lines and boundaries of ethical principles. Religion, therefore, provides a road map; or, in more contemporary terms, religion functions like a GPS to take one on the right path to reach the right destination.

Even before the dawn of religions, morality has taken precedence in all societies through the ages. Even cavemen had to follow a certain code of conduct and observe certain norms to remain part of the clan. Of course, those days required only primitive values, where the needs were more animal than man, as searching for food was the main objective. As Darwin said much later, the survival of the fittest was the guiding rule. Might was the most basic, fundamental and mandatory attribute that everyone had to posses. As the world progressed, changes came in with the prevailing circumstances. The guiding rules of human values changed – as they continue to change as we read this chapter.

The religious code of conduct dictates that ethical values are built within the religious code. Therefore, in order to be ethical, all one has to do is to follow the principles laid down by a religion.

Nusherwan was a Persian king of historical fame as a just ruler, hence called *Adil* or *the just one*, and he was before the times of Islam, as indeed was Hatim Tai who is still today an icon of generosity and charity. But the examples of religiously guided ethical people are inundated with a thousand such folks.

Abraham (Ibrahim AS) is the founder of all three monotheistic religions. Islam is the ultimate religion which brings the final and lasting message to mankind. It not only gives the guidelines for day-to-day life but also saves us from falling into the abyss of evil.

A long time ago, when the civilisation actually took a firm hold upon mankind, man has since endeavoured to identify the good from the evil. It has taken a very long time and the process has not finished yet – not by any standard or any means and may never be; but man has tried, and tried hard over the measurable times and beyond.

We may call the first millennium of the recorded times to be the era of philosophers, the second of the establishment of the religions and faith, and the present millennium may safely be called the age of communication and advanced scientific discoveries. The rapid advancement in knowledge and technology has made it absolutely essential to identify the ethical principles or else the world may fall prey to evil geniuses as has indeed happened in the past.

Ancient civilisations like the Assyrians, the Incas, the Aztecs and, later on, the mighty Chinese dynasties such as the famous Ming dynasty etc., had their codes of conduct. Most of these civilisations practiced authoritarian ways of governance as indeed day-to-day civic lives. The patriarch usually had the authority to decide what was good or bad for the clan or the followers. The chief was often a hereditary owner of the authority, the might and the glory that was part and parcel of the title. He therefore treated his subjects with total authority and without any consideration of their views or thoughts. He was sometimes a god and sometimes a shadow of god as was the title given later on in the human history to the Mughal kings of India; i.e. *Zillae Illahi*. The Incas and the Aztecs worshiped the king, as was often the practice seen later on in the times of Pharoes in the valley of Nile. So who could challenge the god!

The Chinese civilisation is mighty ancient. They had similar code of conduct, as the king was the son of god if not god himself. He ruled his subjects with total authority and tyranny. The tradition of total submission to the kings authority could be seen as recently as the long march by Mao tse dang, who was no less a dictator than the ancient kings beyond the wall of China.

The Indus civilisation was almost contemporary to the Pharonic civilisation of the valley of the Nile. Its remains are still visible in Kotdiji, MoenJo Darro and Harrapa. They had a way of life which was based upon hierarchy, authority and totalitarianism. So was the Pharonic civilisation, which is quite evident from the tales of Moses and his challenge to the Pharro, possibly Ramses the second. Both these civilisations had strong ethical norms, some of which can still be witnessed in the Indo Muslim civilisation of Indo-Pakistan and the present-day Egypt respectively. By the river Euphrates, another massive civilisation was flourishing in the ancient days. It was the Babylonian civilisation, which is still remembered through its founders and eminent rulers like Hammurabi and Nebuchadnezzar. The hanging gardens of Babylon are well known to any student of history. These folks not just ruled the regions, they indeed left considerable indelible imprints of their footprints on human history.

Karen Armstrong is a formidable contemporary writer, historiographer and eminent scholar. In her book named '*Jerusalem*' she mentions that the word *Babylon* is actually 'Bab-Ilani,' meaning 'The gate of the gods.' In Arabic and perhaps in the ancient language, Bab means the gate. Hammurabi was the most famous, benevolent and generous ruler of these lands. Hammurabi also gave the code of conduct which is a lasting gift. Karen Armstrong has mentioned its epilogue in this book as follows:

I made the people rest in friendly habitation;
I did not let them have anyone to terrorise them…
So I became the beneficent shepherd whose sceptre is righteous;
My benign shadow is spread over the city,
In my bosom I carried the people of Sumer, Akkad;
They prospered under my protection;
I have governed them in peace;
I have sheltered them in my strength.
(Karen Armstrong: *Jerusalem*. Ballantyne books. New York. 2005. P19)

Assyrians were responsible for uniting the Near East, according to the French historian Fernand Beraudel. Their unity was further augmented and cemented by the Persian kings, mainly Cyrus the great and Darius etc. (521-485 BC). This symbiotic relationship was shattered by Alexander the Great alias Zulqarnain (336-23 BC) about two centuries later. The massive onslaught by the famous Macedonian gave enormous control over a vast area roughly extending between the Mediterranean Sea and the Indian ocean. They destroyed towns and habitations, but they also established new towns and cities, like Antioch and Alexandria, Syria and Macedonia that produced no less infamous a man than Ptolemy.

The Alexandrian Greeks were more physical than cerebral, akin to the Athenians and the Spartans. The former were philosophers and teachers, the later physical and brawny.

The Romans who followed the Greeks into Asia Minor and the Near East, mainly harnessed the ethical values, code of conduct and the behaviours from the Greeks.

They believed in conquering the world to subdue mankind, enslave them, torture them and plunder the wealth of the world. It was about three centuries later after Christianity had taken hold of the Romans that the Byzantium availed of the fruits of the Greek civilisation. Byzantium prospered and promoted the cultivated values of the Greek philosophers, which partly spilled over into the neighbouring states like Palestine and Syria.

The Near East was also colonised by the Parthian Arsacidae dynasty from around 256 BC and later on by Sassanids of Persia from 224 AD who ruled not just Iran but the huge lands right across the Euphrates and a thousand tributaries; right across the Himalayas into the vast hinterlands of Indus to the Syrian territories. These Sassanids were Zoroastrians and had an established code of conduct, and gave tough fights to the Romans and the Byzantinians. Greek philosophers who suffered humiliation and faced Diaspora in the Justinian era found a sanctuary in the capital of Ctesiphon by the Tigris. The Christian heretic persecuted by the Byzantium rulers reached China through Persia, where they prospered and benefitted mankind.

The Near East was so weakened by the strife and war between the rival forces, that they accepted Christianity. Over time, they were to grow even weaker, so the mighty forces of Arabs conquered them between 632 and 642 AD through Syria, and then Egypt in 639. Persia was to accept Islam soon after in 642 AD. Islam became the driving force in the seventh and eighth centuries as the Arabs and Berbers overran parts of North Africa and finally Spain in 711 AD. In all these vast areas therefore all other ethical principles were replaced by Islamic ethics, as promulgated in the Quran and practice by Prophet Mohammad.

1.2   The Greek Period

The modern journey of ethics began in Athens, where wandering teachers and philosophers like Pythagoras, Thales and Protagoras would teach the mob for a fee. They were called the Sophists and were somewhat condemned by later day philosophers as they charged a tuition fee for their scholarship. Somehow, it is expected of an ethical or a virtuous man to reform the society without a fee! Those were times of huge intellectual growth, and intelligent folks often gathered together in villages, cities and public places to engage in debates, mainly highlighting the virtues of truth, justice, honesty, integrity, etc. Intellectual growth was one of the major objectives of these sophists, even though they were not as respected as they probably deserved.

It was to the credit of Socrates, that the real journey of ethics was duly launched. He was born in a small village in the neighbourhood of Athens and lived for many years teaching Athenians without any financial reward, eventually sacrificing his life to the cause of human rights. Socrates was not a good looking man and as he did not have much of a livelihood in the form of a profession, trade or skills, he was unfairly treated by his wife.

He distributed knowledge to all Athenians free of charge, and without any bias or favouritism. He was a man of sound intellect, great intelligence, and enormous credibility. He would carry his robe on his shoulder, the only worldly asset he possessed and would use it for sleeping on or as a shroud on cold winter nights. He would wander about in the lanes of Athens and gather all and sundry to address not by giving an answer but by raising questions. When challenged that he never volunteered information, instead only asked questions, his answer was simple. He replied that he did not know the answer, hence asked the questions to learn. At one stage he had also said that he knew nothing, and was keen to learn from others.

Once challenged by one of his disciples, that he knew much but did not give the answers, he politely said that he had faith in human intellect and did not want to insult anyone's intelligence and thoughtful exploration of facts by himself; or indeed discourage a person from thoughtful exploration of hidden facts.

The modern way of medical education seems to have reversed back to Socrates as problem-based learning gains popularity throughout the world.

Perhaps the first unethical event took place before the dawn of history, when Iblees refused to bow down to Adam at God's command and Adam ate the forbidden fruit in the Garden of Eden. It was followed by totally unjustified brutality of Cane against his brother Abel.

Justice seems to determine the direction of ethics. It functions as a ship's anchor. Any deviation from justice may result in unethical acts.

'Socrates' was a man of words. He did not leave writing behind. And that appears to be a great shortcoming of the master of all philosophs. Because of this deficiency, we are left with no choice but to benefit of his thought through discussions and debates, narrated by his disciples, particularly Plato, through his famous documents, namely '*The Republic*, *Apology*, *Symposium*, and *Phoedo,*' etc. Plato himself is more famous for his '*Utopia*,' a concept that was more theoretical than practical hence the term '*platonic*' so often used in our day-to-day life. No doubt he was a founding father.

Will Durant has described an episode which is worth a serious thought. He quotes that one day an interesting debate took place in the house of Cephalus – a wealthy Greek aristocrat. Socrates, as we know despised wealth. He asks of Cephalus' "and what do you consider to be the greatest blessing which you have reaped from wealth." Cephalus answers, that wealth is a blessing which enables him to be generous and honest and 'just.' Socrates despite his displeasing demeanour has a sharp and incisive mind. He immediately enquires, "And what sir, is justice." The story goes on to tell us the definition after definition given by one and all who were present in the debate. Of course, Socrates, as we know through history is a difficult customer to convince. No definition pleases him. Annoyed with the stubbornness of Socrates' rather irksome approach, a sophist named Thrasmedus replies with a roar "listen then," he says, to an unfrightened Socrates, "I proclaim that might is right and justice is the interest of the stronger. Different governments make laws, democratic, aristocratic, autocratic etc., with a view to their respective interests; and these laws, so made by them to serve their interests, deliver to their subjects as 'justice' and punish and 'unjust.'"

This doctrine, of course could neither be accepted by Socrates nor indeed by any other philosopher – except Nietzsche who is quoted to have said "verily I laughed many a time over the weaklings who thought themselves good because they had lame paws."

But how ageless that debate proved to be. In most developing nations, after several millennia of that Socratic dialogue, might is not only right but justified to be right. As Churchill once said, that democracy may not be the best form of government, but compared to all other forms, it is still the best. How true! Despite the abuse of democracy in many countries in the year 2013, and the corruption of the very word, its distortion even disfigurement, it is still better than the monarchs of the Arab world, who crush their people with the might and worse, as described in the annals of contemporary history.

Aristotle, who followed Plato, was indeed a sound philosopher. He was a great teacher and a reformer. In the corpus of his works, these are three treatises on ethics. Bertrand Russell, a great philosopher of the past century, in his famous book called the '*History of Western Philosophy*,' raises his doubts on the authenticity of all but one called 'Nicomachean ethics.' In this book Aristotle discusses ethics of length. He believed that the aim of life is "not goodness for its own sake," but "happiness." Further in the argument of defining happiness, Aristotle devises a road to excellence. He believes that such a path will be a path of morality and be based upon ethical values.

The qualities of character, according to Aristotle can be arranged in a triad; in each of which the first and last qualities will be extremes and vices, and the middle quality a virtue – or an excellence.

"So between cowardice and rashness is courage; between scoth and greed is ambition; between humility and pride is modesty; between secrecy and loquacity is honesty; between moroseness and buffoonery is good humour; between quarrelsomeness and flattery – friendship; between Hawlet's indecisiveness and Quixote's impulsiveness is self control."

Describing the virtues of an ideal human being, who is a picture, may be a measure of morality, this great philosopher says.

"He does not expose himself needlessly to danger, since there are few things for which he cares sufficiently; but he is willing, in great crises, to give even his life, knowing that under certain conditions it is not worthwhile to live. He is of a disposition to do men service, though he is ashamed to have a service done to him. To confer a kindness is a mark of superiority; to receive one is a mark of subordination … He does not take part in public displays … He is open in his dislikes and preferences; he talks and acts frankly, because of his contempt for men and things. He is never fired with admiration, since there is nothing great in his eyes. He cannot live in complaisance with others, except it be a friend; complaisance is the characteristic of a slave … He never feels malice, and always forgets and passes over injuries… He is not fond of talking. It is no concern of his that he should be praised, or that others should be blamed. He does not speak evil of others, even of his enemies, unless it be to themselves. His carriage is sedate, his voice deep, his speech measured; he is not given to hurry, for he is concerned about only a few things; he is not prone to vehemence, for he thinks nothing very important. He bears the accidents of life with dignity and grace, making the best of his circumstances like a skilful general who marshals his limited forces with all the strategy of war. He is his own best friend, and takes delight in privacy whereas the man of no virtue or ability is his own worst enemy, and is afraid of solitude." Such is the Superman of Aristotle. A model of morality.

I have read these paragraphs many times, and searched for a person in human history to match the attributes described by Aristotle. Plato was platonic, not Aristotle. He was a practical man, so obviously he expected a real, living, breathing human being to match his criteria. Then I came across a writing of Dr. Ali Shariati, a leftist philosopher of the twentieth century, who was eliminated by Shah of Iran, for the fear of revolution that this man could bring about through his writings.

In his essay called 'Ali, a mythological truth', Dr. Shraiati discusses many virtues that Imam Ali had, finally concluding that to possess the contrasting attributes and do justice to them all is what makes Ali a unique person. He was a philosopher and a soldier, a warrior and a peace lover, a speaker and a writer, a disciplinarian and a gracious ruler, and so on. That is why Imam Ali was a mythological character who lived on this earth in the not so distant past. Dr. Shariati calls him the Prometheus of Islam.

So when I look at the attributes described by Aristotle in an ideal man, they suit and match no one except Imam Ali. That is why people love him so intensely and live and die for him.

Nietzsche also described attributes of an ideal human being which are not much different to Aristotle, though more macho. Iqbal the philosopher poet of the East, harnessed Nitsche's thought and presented his concept of *marde momin*, portraying Imam Ali.

'Myths', as anthropologists tell us, exhibit social pattern and structure. "Myths and rituals together provide a means whereby men can exhibit to themselves the forms of their collective life", writes Alasdair Macintyre in his excellent book called *'A Short History of Ethics.'*

"God is our father. God commands us to obey him. We ought to obey God because he knows what is best for us, and what is best for us is to obey him", writes Macintyre in the chapter on 'Christian Ethics.'

"The first conceives of moral precepts in terms of commandments and of moral goodness is in terms of obedience."

"All religions teach one thing, namely the goodness of deed." To be kind is to be good, hence moral. To be obedient is good, hence moral and to be honest and just are good deeds, hence moral. No religion condones falsehood, treachery, dishonesty, injustice and cruelty. Indeed Judaism as well as Christianity both form the foundation of moral values upon which Islam is based. The major difference being that while Judaism as well as Christianity both became vulnerable to interpretation of the God's books by human beings, Islam remains totally divine through the jealously guarded revelations of Almighty Allah in the Holy Quran – pure and simply original document on ethics. Not a word has changed in the teaching of Quran, as opposed to the Torah and the Bible.

Jesus Christ was a Prophet, who preached morality and honesty, piety and justice. He sacrificed his life for the upholding of truth. Regrettably his followers could not follow him in the true spirit of his commandments.

Therefore, we see that the early Christians, although they claimed to be fair and honest and moral, could hardly justify the incest so blatantly practiced by the Borgia popes.

Three systems of ethics; three ultimate concepts of ideal moral life seem to have dominated the ages. Budha preached and practiced modesty, simplicity, humility and fairness to all. He, like Jesus, considers all men to be equally precious, resisting evil not by reciprocity but by returning good, identifying virtue with love, including democracy in politics to limitless power of the people.

Islam is one religion that overshadows all other heavenly or man-made doctrines, in preaching the goodness of good over evil, but awarding punishment to the evil-doer. Forgiving is better than avenging, and yet Islam insists upon punishing the bad deed. Hence, while Christianity proclaims offering the other cheek, Islam commands to amputate the hand of a thief or award death penalty to a murderer.

Therefore, we see that Budha as well as Jesus, both reformers in their own rights, preferred softer virtues, as opposed to Islam which prefers courage, boldness, firmness, steadfastness and strength.

A second system of ethical principles to have dominated the era, at least of the European region is that of Machiavelli and Nietzsche. They both stress the masculine virtues, accepting the inequality of men, enjoy the pleasure of victory in combat, relish the sanctity of martyrdom of the slain, relish the pleasure of conquest and authority. They identify virtue with power, and believe in hereditary aristocracy.

Of all the great philosophers of the universe, ethics of Socrates, Plato and Aristotle must be given the room at the top. These super masters deny the universal acceptability of either the feminine or the masculine virtues. They consider that only the informed and the mature mind can judge, according to diverse circumstances, as to when love should rule and when the power be utilised. Their ethics identify virtue with intelligence, mixing aristocracy with democracy in suitable proportions. In their view, knowledge, wisdom and foresight based on faith 'Yaqeen' in one's cause form the inner core of 'ethics.'

Perhaps Islam is one religion that gives more importance to justice than any other philosophy. It appears that the ethics of Socrates, Plato and Aristotle duly displayed in the teaching of Islam rather overtly could raise very basic questions in the minds of his audience who were mostly young people, such as democracy, freedom of thought and expression, evils of authoritarianism, oligarichism or tyranny, etc. He would stimulate the minds of these people by simply discussing basic facts without exaggerating or diluting the facts. He was the finest teacher ever, but did not dictate nor indeed personally document his thoughts. He believed that human mind should be cultivated to memorise and recall when needed. It was by itself, a major handicap in the life and times of this great philosopher. Documentation was discouraged as he considered it an insult to human mind to note down rather than memorise.

Islam has bridged the gap between ancient times and modern days through its massive input in the philosophy of ethics. And indeed by transferring the knowledge gathered by such greats as Plato and Aristotle, to the West.

Of course the greatest of all philosophers, reformers and thinkers, the Holy Prophet (PBUH) himself taught his followers the virtues of goodness over evil, condemned tyranny, supported the cause of the masses, promoted education, fought against ignorance and brought discipline into the lives of the pagan Arabs. The most illustrious seat of learning law and justice in the world called the 'Lincoln's Inn', stands adjacent to my alma mater, the Royal college of Surgeons, only a few minutes walk from the most renowned London School of Economics, not far from High Holborn. At the top of the list of the greatest law givers in the world shines the name of Prophet Mohammad.

One witnesses through history that the Holy Prophet was tortured and tormented by his tribesmen; but he tolerated all with great patience, only to reform them.

Ethics of Islam excels all others teachings. Let us investigate this hypothesis.

In a chapter entitled 'The ideals of Islam' in the famous book '*The spirit of Islam*,' renowned historian of the Indian subcontinent Syed Ameer Ali writes, "The religion of Jesus bears the name of Christianity derived from his designation of Christ; that of Moses and of Budha are known by the respective names of their teacher. The religion of Mohammed alone has a distinctive appellation. It is Islam."

Salam means peace and Islam means total submission to God.

Amongst innumerable '*Sifaat-e-Illahi*' the mightiest of Lords, the most oft repeated are '*Al-Rehman* the Benevolent' and '*Al-Adil*', i.e. the just. No religion or moral philosophy attaches more significance to justice and benevolence than Islam.

Syed Ameer Ali writes, "The essence of the ethical principles involved and embodied in Islam is summarised in the second chapter of the *Quran*. There is no doubt in this book." A.L.M., 'Zalika – al-Kitab, La-Reiba' Feeh – "a guidance to the pious, who believe in the unseen, who observe the prayers, and distribute charity out of what we have bestowed on them, and who believe in that which we have commissioned them with, and in that we commissioned others with before thee and who have assurance in the life to come. Here I have received the direction of thy Lord." *(Quran, Sura ii- 1–6)*.

The principle bases on which the Islamic system is founded are (1) a belief in the unity, immateriality, power, mercy and supreme love of the Creator; (2) charity and brotherhood among mankind; (3) subjugation of the passions; (4) the outpouring of a grateful heart to the giver of all good; (5) accountability, for human actions in another existence.

Islam grants more importance to quality of mankind, justice and fair play than either Christianity or Judaism. The Prophet of Islam was called *Al-Ameen*, the trusty even by the worst of his enemies. His entire life lay exposed before the Pagan Arabs. He was born and brought up amongst his kinsmen, so they could not say that any aspect of his life remained occult. The Prophet SA displayed through his deeds and actions that Islam attaches, huge importance to the freedom of mankind, honesty, trust, faithfulness, humility, piety, justice and charity.

'Prayers' are mandatory to inculcate discipline and subjugation to the creator, the master, and the Lord of life and death. 'Fasting' is ordained to appease one's worldly desires, depriving oneself despite the availability. The fast also teaches tolerance, pain, discomfort and agony of not receiving – thus realising the feelings of the 'have nots.' And '*Zakat*' (alms), an essential pillar of faith displays the real soul of Islam. To give away is Godly virtue, to receive is human. By giving '*Zakat*' one displays the reflection of God unto oneself. To part with your possessions demands the ownership of spirit of sacrifice. The outcome of this sacrifice is the pleasure of witnessing one's neighbourly folks at par with oneself.

Islamic code of ethics is further highlighted in the fourth sura of the Holy *Quran*. "Come I will rehearse what your Lord hath enjoined on you, that ye assign not to him a partner, that ye be good to your parents; and that ye slay not your children because of fear, for them and for you will we provide; and that ye come not near the pollutions, outward or inward; and ye slay not a soul whom God hath forbidden, unless by right, and draw not might to the wealth of the orphan, save so as to better it – and when ye pronounce judgment then be just, though it be the affair of a kinsman. And God's *compact* fulfil ye; that is what, what He hath ordained to you. Verily this is my right way follow it then." (*Quran*; sura iv: 155).

And further on "Blessed are they who believe and humbly offer their thanksgiving to their Lord … who are constant in their charity, and who guard chastity and who observe their trust and covenants … verily God bids you do justice and good, and give to kindred their due; and He forbids you to sin and to do wrong and oppress" (Al-*Quran*; sura iv).

Thus we witness through the actions of the '*Moalam-e-akhlaq*' the teacher of Morals, the Prophet of Islam (PBUH) through his life as well as his last sermon that he gave away a charter of human rights. He granted freedom to slaves, forgave the murderers of his close relatives, strongly urged equality, parity and justice; barred all differences of colour, race or creed; and duly commanded everyone to obey the Almighty so that the rightful path be followed. He commanded every follower to acquire education, as the darkness of ignorance could be dissipated only through knowledge. A knowledgeable person can find his path differentiating good from evil.

History is full of episodes and illustrations of the deeds and actions of the companions of the Holy Prophet (PBUH), and his progeny. They not only followed each and every word they learned from the Holy *Quran* but strictly observed the practices of the Holy Prophet (the Sunnah).

1.3 The Arab and Muslim Period

It is wrong to say that the world plunged into the abyss of darkness after the demise of Athens, until the renaissance in the fifteenth century. It simply means that many western writers are either ignorant or in denial of the contribution made by Islam during the dark ages of Europe. Let us therefore investigate the facts.

The Arabian Peninsula was immensely submerged in most evil culture until Prophet Mohammad was born in Mecca in April 571 AD He was born a prophet but waited for 40 years to declare his prophet hood. During all those years he led an exemplary life, so that people could watch and observe him critically and minutely. His credentials were finally established as the *Ameen* and Al-*Sadiq*. He presented his doctrine of ethics as dictated by Allah ST through the *vahi*. He preached for 23 years before returning to his creator at the age of 63 years. His life and times are well documented in the annals of human and Islamic history. The principles of Islamic ethics are discussed elsewhere in this book.

The primary sources of Islamic ethics are: Quran and Sunnah. Followed by logic, ijma of early scholars, and Ijtehad; which is a way of deriving laws from the rightful sources.

Islam is a religion based upon philosophy. The moral code of Islam is dictated in the *Quran* as '*Amr Bil Maroof wa Nahi Analmunkar*' roughly translated as 'bid good and forbid evil.' It was practiced and preached by Prophet SA and his family right to the core of its content. Imam Ali has been aptly described by the western historians as the philosopher soldier of Islam. He indeed was the foremost teacher of ethics, *ilam* o *idrak* not just in Medina for a quarter of a century after the Prophet departed, and Ali faced social boycott by the Muslim *umah*, devoting 25 years of this part of his life in social work for the community. He dug wells, planted date palm orchards and built mosques all along the path from Syria to Mecca. For centuries the Hajjis benefited with these facilities.

The moral code mentioned above was also the mission statement of Imam Hussain AS in Karbala was, as declared by him in many of his sermons. Imam Zain Al Abedin has written a *dua* in his collection called *Sahifa* e *Kamila*. It is called '*Duae Makaram Al Akhlaq*' This *dua* defines and describes the cardinal principles of ethics; both normative as well as applied; narrated in a fashion that only an Imam could do.

Between the seventh and the beginning of the fifteenth century Muslim philosophers, scholars, scientists, teachers, reformers and learned men dominated the world. They mastered not just humanities, philosophy, logic and literature but also produced great mathematicians, astrologists, physicists and chemists. Historians have named this era as the 'golden age' of mankind. These centuries infact bridge the gap between the decline and fall of the ancient Greek civilisation and its great philosophers and the renaissance of the West after the fifteenth century. The contribution of Islamic philosophers and scientist is somehow underplayed by the West and the period between the disappearance of Greeks and the renaissance of Europe as the dark ages. It is totally false. This was the golden period of Islam, during which knowledge of ethics, good and evil code of conduct in all matters of life and death and indeed a thousand other elements were taught by the followers of Prophet Mohammad SA. They not only taught through Islamic teachings but also transmitted the teachings of Aristotle et al. through the translations, particularly during the reign of Mamun Rasheed, a student and disciple and finally the murderer of Imam Ali Al Riza the great grandson of Prophet. This material was then transferred through many great institutions and scholars to the rest of the world. And the Fatimid caliph Al-Hakeem established the illustrious university and named it after Hazrat Fatima ut Zahra as as Jama-al Zahra. It continues to serve humanity through promotion of knowledge, though, once the Kurd Salahuddin Ayubi destroyed the Fatimid dynasty; the later rulers converted the name of this university to Jama-al Zahra, a great source of ethical teaching in contemporary world. Earlier on during the Fatimid's rule in Egypt, the most renowned university called Jama al Zahra. was established by Al Hakeem. It was named so, after the daughter of Prophet, as the Fatimid's traced their descent from Hazrat Fatima SA. It was primarily designed to teach Islamic *fiqh* and *hadith*, *ilme kalam* and theology in the light of the teaching of Shia thought. A few centuries later, after the down fall of Fatmides at the hands of Salah Ulddin the Kurd alias Ayubi, the university was renamed as Jama Al Azhar. It continues to spread knowledge throughout the world a thousand years after its creation.

The tragedy and massacre as well as annihilation of prophet's household at Karbala occurred in Higra. For nearly three decades the household of the Prophet was actively persecuted in Medina after the Karbala, and Imam Zainal Abedin was totally deprived of any contact with his followers. It was Imam Baqar who took up the responsibility of continuing the process of spreading education to the masses through establishment of numerous schools in and around Medina. It is quoted in the literature that at the end of his career as a teacher and a philosopher he had educated thousands of students in Fiqh, logic, theology, philosophy, astrology, mathematics, diction, geography. Some of his students came from India, possibly from Sindh. Imam Zain al Abedin had a wife from Sindh named Jawaraiah or Huriah from Thatta. So the links had been established between the household of the prophet and India. Hence the cousin of Imam Baqar lived in exile and hiding from ummayads in and around Thatta for several years, and was finally murdered by Hisham bin Malik. He was Abdullah Al Ashtar also called Abdullah Shah Ghazi, the grandson of Fatimah Kubra and Hassan al Mussana; who was buried on an island in the Arabian ocean, now called Clifton, in Karachi. For more than a thousand years, this saint and martyr of the family of prophet continues to spread the message of peace and be a source of hundreds of thousands of poor and hungry folks through charity offered at his mausoleum. The greatest human tragedy of all times in the form of brutal massacre of Imam Hussain and family In Medina Imam Al Baqir established a school for the education of common man. In this school he taught languages, mathematics, humanities, logic, history, diction, debate, fiqh and habit. Copies of the discourses conducted in his institutions are available in many libraries in today's world. They are simply mind boggling.

His illustrious son Imam Jafar Al Sadiq was the only descendant of Prophet who could find some years of respite as during his early days, Abbasids were demolishing the Umayyad, and had no time for putting the Imam in the dungeons as the later day monarchs of this dynasty did to the later Imams.

Imam Jafar Al Sadiq (82H or 702-765 AD) had over 4,000 students in his university at Medina. He had thousands of students across the lands of the Muslim umah and elsewhere. Amongst his famous students were men like Imam Abu Hanifa the founder of Fiqha hanafi and Imam Malik Ibne Ans the founder of Maliki fiqh. Jaber Ibne Haayan, the father of chemistry called Geber in the west learned his chemistry at Imam's feet and dedicated his works 'to my master, a mine of wisdom.' His books were taught until the sixteenth century in Europe.

Literature search informs us that amongst the guiding philosophers and Faqeehs of Islam, Imam Jafar Al-Sadiq just like his father and grandfather is the first and the foremost philosopher and scientist. His students include Imam Abu Hanifa, Malik ibne Anas, and Wasil bin Ata, etc. Literature search also includes the names of

Imam Bukhari, Imam Muslim, Al Tirmidhi, Ibne Maja, Abu Daud and Al Nasa'I as learned teachers of Islamic jurisprudence.

Amongst the golden philosophers of Islam, following names top the list. They are:

Al Kindi (D in 873 AD)

Al Farabi (B in 870 and D in 950)

Abu Ali ibne Sina (B 980–1037)

Al Ghazali (D in 1111)

Ibne Rushd (B 1126 D 1198)

In the chronological order if we look at the appearance of philosophers in Islam, we begin with Prophet himself. He departed in 632 AD.

Ali Ibne Abi Talib left us three decades later, but he was not allowed by Umah to benefit them with his knowledge as he so much wanted to and lamented at it in his letters and lectures recorded in Nahaj al Balagha.

After the great tragedy of Karbala, an important watershed in Islamic history, Imam Baqir began the service to the cause of education. He lived from 676–733 AD. He was followed by Imam Al Sadiq the teacher of all other ensuing philosophers. Imam Al Sadiq lived from 702–765 AD. In 750 AD, Abbasids massacred Umayyad and confiscated the caliphate in the name of the household of Prophet. The turmoil of unprecedented nature did not allow Imam Musa Al Kazim any respite. He spent most of his life in the Abbasid dungeons.

As mentioned before, amongst the many students of Imam Al Sadiq was a man called Wasil bin Ata. He was rusticated by Imam and later on introduced his own theological principles. His followers were called the Mu'tazzalites. They believed that the *Quran* was a word of God but written by man, so tahreef was a possibility. Abbasid kings promoted his cause to bring about further disgrace to the Prophet and his message.

Al Kindi was a famous Muslim scientist and philosopher who died in 873, his father was governor of Koofa, and was contemporary of Al Razi, and was often called the philosopher of the Arabs.

This great Muslim philosopher whose name was Abu Bakr Muhammad Bin Zakkaraiyah al Razi (Rhazes 865-925 AD) was a contemporary of Al- Kindi His famous treatise called *Hadi al Attiba* (a guide to the physicians) was translated by the Jewish scholars into Hebrew and contained several writings of Islamic Medical ethics. Here is a quote from Al- Razis' book:

"A physician should have full faith in the art of healing. He must never try to achieve greatness or honour by disgracing or humiliating others. He must remain ready to serve the poor and the needy patients at all times and treat them with efficiency as it is a noble act of service to mankind. Even though he may not be sure of the recovery of the patient from illness or disease, yet he must continue to infuse confidence for an early recovery. Thus he will assist the patient's natural power of recovery (vis medicatrix naturae)."

Al Razi also wrote another book called *Al Tibb al Ruhani*, which dealt with both normative and applied ethics in the care of the patients. It was a compliment to Al Razis famous book called *Kitab Al Mansuri* based on his medical knowledge, skills and experience. Al Razi was the chief of a hospital in Ray near Baghdad, and wrote five invaluable books on medical ethics. His famous instruction to his students was "As long as you can treat patient with diet alone, do not start him on drugs. If the drugs are to be given, always begin with the simplest drug first, and proceed to more complex ones later … If the physician is knowledgeable and skilful and patient compliant, the disease may not find it easy to progress further. Treat your patient so that he does not lose his resistance (immunity) both mental and physical."

Al Farabi was born in 870 AD. He was Turkish, lived in Aleppo and died in Damascus. He was also called the 'philosopher of the Arabs.' It is said that he was also involved in research that at his death bed he questioned his visitor about the puzzle they had discussed a few weeks ago. When the visitor, obviously himself a man of letters, said "but Al Farabi you do not need the answer as you are on your way to meet your creator." To which it is said that Al- Farabi turned his face away from the visitor muttering; "alas, even you my friend, want me to die in ignorance!"

Abu Ali Husayn bin Abdullah ibne Sina alias Avicenna was a great physician, philosopher and ethicist of all times. He laid down the edifice of research in medicine through practical experimentation and careful observation physician should be true to. He observed and maintained great respect to the sanctity of life and *documented his works in his book. He wrote the most illustrious Qanoon Fil Tibb alias the Canon* of Medicine which was employed by the Muslim universities for numerous centuries as indeed by the European universities of medicine until about the sixteenth century.

He advised that a physician should have firm faith in Islam and its teachings. He should be honest to himself, his profession and above all to his patient. Abu Ali Sina kept himself in line with the Mutakallimun (dialecticians). To him Allah is the ultimate healer, but a physician must serve his patient to the best of his ability, with honesty and integrity.

Al Ghazali was a radical thinker. His work called '*Tauhafat al Falasifa*' or 'the incoherence of the philosophers,' made the philosophers look silly. But he in fact meant to critically examine the contemporary philosophers and improve their thought. His Sufi tradition has spread in the Islamic world and has many followers. (830-870AD)

Ali ibne Rabban al Tabari was a teacher of Al-Razi. One can therefore imagine the grandeur of the knowledge and expertise this man must have possessed. His book called the '*Firdaus al Hikmat*' is a medical encyclopedia. It is comprised of seven parts. The first treatise deals with Kuliyat e tibb; i.e. the principles of medicine. In it he wrote:

It is essential that a physician be kind, generous, and content and possesses a good character. In fact he should be kinder to patients than to his kith and kin; and put the patient's welfare above his self interest. He should act rather than simply talk. Neither should he be haughty nor indeed greedy for money; as these evil traits will damage his high ranking status that society has given him, his prestige and his profession. The physician must use only those medicines that have been tried and tested.

Ibn Rushd is perhaps the most talked about philosopher in the West. He is known here as Averroes. He lived in Muslim Spain and was a contemporary of the famous Jewish philosopher Moses Maimonides. Ibne Rushd did not differ in his theory from some other greats like Ibne Bajjah and Iben Tuffail who in turn followed the philosophy of Ibne Sina and Al Farabi.

During the Muslim rule of Spain, many other Jewish philosophers like Ibne Gibbons, Narbonne, etc. flourished and translated the Arabic philosophers works into Hebrew. After the fall of Grenada, in 1492, described by the Christian crusaders as 'Reconquesta,' these Jewish lovers of knowledge went into hiding. They left Spain to re-emerge in many parts of Europe in the form of Diaspora, and transferred the works of Arab philosophers to many European philosophers in the fifteenth and the subsequent centuries.

Ibne Rushd left a lasting impression on the thinking of one of the greatest European philosophers by the name of Thomas Acquinos. It appears that Spinoza and Descartes were influenced by Thomas Aquinas and therefore indirectly with the philosophy of Ibne Rushd.

In the ninth century, Ishim bin Ali Rahawi wrote a book called '*adab-al Tibb*' which was perhaps the first documented treatise on medical ethics. He wrote on many topics covering a wide range of subjects related to the patients' rights and physicians' obligations. Besides Al Razi and Ibne Sina, this philosopher concentrated hugely on the principles and parameters of medical ethics. Ibne Nafis of Egypt described blood circulation in the human body several centuries before William Harvey, and documented his works along with the ethical principles of research he employed in his book.

The role of ulema in the propagation of knowledge has always been pivotal. The history is full of men of immense gravity who served the cause of education throughout the Muslim ear and have continued to do so since the times of early Hawzahs established in Najjaf al Ahraf and later on Qumm.

Just to quote a couple of examples, one may look at the life and times of Abu Jafafr Muhammad ibn Muhammad, Khawaja Nuruddin Tusi (D 672/1274) He was an acclaimed astronomer and mathematician. He also wrote on medicine, ethics, history and geography.

Another alim of singular importance was Jafar ibnal Hasan Muhaaqqiq al Hilli, also called Muhaqqaq al Awal (D 676/1277). His nephew Hasan bin Yousef Allama al Hilli (D 726/1325) was one of the most influential shai ulema of the middle ages. Allama Hilli was responsible for introducing *Ijtehad* as the core method of Shia Islamic theology [18]

1.4   Bacon: Russell-Mcindyre-Dawkins, Beauchamp and Childress

Francis Bacon was called one of the founding fathers of modern western philosophers, by no less a person than Bertrand Russell, who was himself not only a great philosopher, but also a top mathematician and nobel laureate of the twentieth century. Bacon was credited by Russell as an introducer of inductive method and a pioneer of logical systemisation of the scientific procedures. Bacon is also given the credit for approaching the problem based upon logic rather than mere hypothetical deductions. He discussed masterly on sensitive topics like faith and truth in all day-to-day matters and reached definitive answers through logic. He also believed in the principle of 'double truth' that of 'reason' and 'revelation.' Bacon advocated the cause of reason in deducing philosophical principles and never compromise on the principle of truth.

Spinoza was one of the greatest Jewish philosophers who was the by product of Jewish Diaspora after the Crusaders expelled Muslims and their subordinate Jews from Andalusia. Jews had to migrate from Tyre, Sidon and Jerusalem, and from Athens to Antioch, Alexandria and Carthage – some finally settling down in the Moorish Spain, until 1492. The miserable and migrating Jews eventually found some respite in Holland. During their renaissance, they suffered even more but resolved to peruse the independent professions through acquiring finest education, producing philosophers, physicians and above all economists and controller of finances.

Spinoza was born to such migrating parents in 1632. Will Durant describes Spinoza as the child of exodus. Three pivotal terms were used by Spinoza namely 'substance' attribute 'and mode.' Durant says that Spinoza's ethics flows from meta-ethics. His thoughts complex, beguiling and intriguing, but full of wisdom, knowledge and pursuit of truth.

Another great philosopher who changed the flow of human history was Voltaire. He was ugly, flippant, vain, obscene, unscrupulous, even dishonest in the eyes of historian Durant. But Voltaire was Voltaire. He alone changed the French history through his powerful writings sensitising the masses, awakening in them the love of democracy and dislike for monarchy. He alone brought in the French revolution beginning in Bastille and spreading through the cities and towns of France, bringing in one of the finest versions of modern democracy. Once questioned about his trade, he answered "my trade is to say what I think." And he certainly was deep thinker, and a fine wordsmith too.

Voltaire was arrested 11 times and exiled from France, for truth telling and showing the right path to freedom and human rights for his people. Voltaire believed that the human mind could not be liberated without knowledge. He also wrote in his works that truth eventually prevails over falsehood, and honesty over use of unfair means. Once my teacher, the illustrious Professor Karrar Hussain, pointed out to me that had it not been due to Voltaire, Europe would still be in the clutches of tyrant dynasties.

Voltaire wrote; "History is nothing more than a picture of crimes and misfortunes." History, he concluded, "is after all nothing but a pack of tricks which we play upon the dead. We transform the past to suit our wishes for the future, and in upshot 'history proves that anything can be proved by history.'"

His greatest book on philosophy of history brought him exile. He told the truth, thus enraging the clergy that the rapid conquest of paganism by Christianity had disintegrated Rome from within to encourage barbarians – leading to the fall of the Roman empire. He gave even less importance to Judaism and Christianity, and more to China, India and Persia. Finally, the king decided that this man who could dare to think of himself as a man first and a Frenchman after wards should never put a foot upon the soil of France again.

According to Will Durant, Voltaire believed that human mind could not evolve without knowledge. Truth prevails upon falsehood and honesty upon unfair means, lies tell a story of the 'good Brahmin' who says, "I wish I had never been born." "Why!" so I said. "Because", he replied, "I have been studying these 40 years, and I find that it has been so much time lost. I believe that I am composed of matter, but I have never been able to satisfy myself what it is that produces thought. I am even ignorant whether my understanding is a simple faculty like that of walking or digesting, or if I think with my head in the same manner as I take hold of a thing with my hands ... I talk a great deal, and when I have done speaking I remain confounded and ashamed of what I have said."

"The same day I had a conversation with an old woman, his neighbour. I asked her if she had ever been unhappy for not understanding how her soul was made? She did not even comprehend my question. She had not, for the briefest moment in her life, had a thought about these subjects with which the good Brahmin had so tormented himself. She believed in the bottom of her heart in the metamorphoses of Vishnu, and provided she could get some of the sacred water of the Ganges in which to make her ablutions, she thought herself the happiest of women. Struck with the happiness of this poor creature, I returned to my philosopher, whom I thus addressed; 'Are you not ashamed to be thus miserable when, not 50 yards from you, there is an old automaton who thinks of nothing and lives contented?' 'You are right,' he replied. 'I have said to myself a thousand times that I should be happy if I were but as ignorant as my old neighbour; and yet it is a happiness which I do not desire.'"

Theory of evolution presented by Darwin had created many ripples in the nineteenth century. Resulting in revolutionary theories.

Enlightenment was a great watershed in European history. So much happened after this landmark period that not just Europe, but the whole world changed.

Kant believed in God, freedom and immortality. God is required as a power, supreme and ultimate, capable of providing, giving awarding or punishing. Freedom is required, because happiness is a virtue which can only be achieved through happiness. Immortality is required because virtue and happiness manifestly do not coincide in this life.

Hegel was also a German philosopher who is considered to be the final post in the field of ethics. His whole philosophy is an attempt to show that the history of philosophy is at the code of philosophy.

In his famous book the '*Philosophy of History*', Hegel outlines the concept of master-serf relationship by Karl Marx and his theory of class struggle.

Karl Marx was a son of Jewish parents who lived in Germany but due to sociopolitical reasons they adopted Christianity. He along with Fredirch Engels is the father of the twentieth century drama called 'Marxism.'

His philosophy flourished for approximately 50 years in the Soviet Union, finally rejected by the contemporary statesmen, Gorbacheve and Yeltsin.

Marx believed that religion was the opium for the masses. He believed that in the greed of achieving an abode in paradise in the life hereafter (if any), men forget to take care of their basic rights in day-to-day life. For them, religion functions as a means of flight from reality. He believed that if with a struggle one could achieve all the comforts of life promised by religions in the life hereafter, they would not care much for religion. Religion would thus die its own natural death.

Marx considered religion to be the cause of immorality, a relic of the past, memorabilia of the bygone days.

Marx did not believe in the lasting moral values. He believed that with the change in the social set-up, the moral values also change. He believed that just as the industrial revolution debased the moral values of the preceding aristocratic values, a socialist revolution shall destroy the ethical values of the machine age. And the process shall go on.

Marxist philosophy propagated the cause of the poor and the downtrodden. It supported the notion that if a worker does not have to worry about the livelihood of his family, he would not be influenced toward amoral acts. A sense of security inculcated by a social set-up as presented in Marxism would automatically prevent any amoral, unethical action of an individual indeed a society.

Theoretically, of course the Marxist philosophy appeals to a common man, but despite its unprecedented support and popularity in the twentieth century following the first great war, the ultimate fate of socialism has not been much different from other preceding philosophies. What has happened in Soviet Union, where free market economy system has destroyed the socialist philosophy, may be a lesson in history.

While God's laws in the form of Judaism, Christianity or Islam have lasted thousands of years whereas the man-made philosophy has not even lasted a full century!

Bertrand Russell is a household name of the recent times. Indeed he is one philosopher who received more recognition during his lifetime than anyone else amongst his predecessors. He was born and lived in an electronic age, confronted major moral issues; e.g. genocide, atomic war and other war-games with great courage and considerable success.

Russell was primarily a philosopher mathematician who believed in the 'rigid impersonality,' and objectively of the subject and called it the 'eternal truth.'

He once wrote an essay 'why I am not Christian,' in which he claimed that all religions teach one thing, i.e. supremacy of truth, justice and good deeds.'

So he argued why should he hurt the feelings of a Muslim by becoming a Christian, or for that matter a Buddhist or a Hindu, etc.

Russell wrote, "The free man cannot comfort himself with childless hopes and anthropomorphic gods; he has to keep his courage up, even though he knows even he has to die."

After the great war, a radical change occurred in Russell's philosophy. He considered the state to be responsible for the manslaughter and genocide. What good could come out of such a collective madness called 'state' he argued. Freedom he thought was the supreme good. Life and knowledge are so complex in today's world, that only free thinking and dialogue could isolate good from the evil.

Russell gave unprecedented importance to knowledge, speech and the freedom to speak. He argued that knowledge alone could train the mind to outline the parameters of goodness, sifting the evil out. He argued the distinctive feature of the unintelligent man is the hastiness and absoluteness of his opinions; while a knowledgeable man – a scientist is slow to believe, and never speaks without modification.

Russell was awarded a Nobel prize for peace and has only passed away in the last century.

Xuan Paul Sartre was born in 1905, reaching the pinnacle of glory receiving many accolades in our own time. He was agnostic at an early age and an atheist soon after. He was a product of wartime France. He opposed tyranny, the Nazis and their inhuman activities almost throughout the period of war. His famous book called 'Being and Nothingness' denies the existence of many human norms. In his heart he was a pacifist and theosophical, but very lonely and generally an unhappy character.

He attaches more importance to freedom of expression than even Russell. Sartre's moral code is very confusing. He is a pessimist who calls 'freedom' a liability on the human shoulders, He writes:

"My freedom is the basis of my ethical code, because my existence is the cause of such codes. Therefore, it does not behove upon me that I should follow one or adopt another moral code. In view of the existence of values, which form the very core of human life, my existence is of no consequence. And my freedom is doomed to learn that freedom itself has no basis, is meaningless and worthless." Sartre died a sad man, bitterly failing to find peace unto himself.

1.4.1 The God Particle

Richard Dawkins is a contemporary philosopher who is engaged in disproving the existence of God. He is an extremely articulate thinker, speaker and writer. His famous book called the 'Selfish Genes' brought him huge accolades and worldwide recognition. His latest book called 'Delusion of God' is an extremely impressive book to read. Dawkins leaves no stone unturned to disprove the existence of any form of deity. He claims that nearly all the famous scientists from Einstein to Stephen Hawkins are nonbelievers, and are still looking for an entity that would convince them about the presence of an ultimate power, the creator of life.

They should all be delighted to learn that in the first week of July 2012, an illustrious group of scientists working on an underground equipment of huge proportions called the Dihydron Collider or something similar, sitting underneath the border of France and Switzerland have finally found their god, which they have described as 'Higgs Boson' named after Peter Higgs and Satyaynadar Nath Bos, They had predicted decades ago that there must be an ultimata infinite energy particle possessing 'mass as well infinite energy' and have omnipotent capacity to generate, regenerate or create electronic activity indeed life itself, at the most minuscule nano level. In fact, Peter Higgs gave an interview on British television that he had predicted the presence of an infinite Boson, but did not expect it to be discovered in his life time. Nuclear scientists had already found out the atom is much too large an entity and has further components like proton, neutron and electron. In fact, even the so-called finest particle, i.e. quartz has finer components called Bosons. They knew all that until they found out the missing link, which they call the 'god particle.'

Anjum Niazi is an excellent columnist. In her column in 'Daily Dawn,' a renowned Pakistani newspaper, she lamented on the loss at the hypocrisy of Pakistan that the only Nobel Laurette of Pakistan, who actually won the said prize on the particulate theoretical physics half a century before Higgs and Bos, that indeed was the most fundamental discovery leading to the latest achievement of science. Just because Dr. Salam had a different faith than the majority, Pakistanis hate him, disown him and have eradicated even his name from the history books.

Thank God they have found the infinite particle that they may call whatever they wish. One hopes that now they would join the right path that all religions have taught.

For a believer, the greatest consolation is that there is only one God, who has created everything in the universe and the cosmos. He alone is the creator and to Him alone shall every creation ultimately return. Even those who have found the Higgs Boson!

The human history is full of epic tales of bravery, valour, justice, honesty, integrity and many other attributes that will fall in the category of virtues. As mentioned before it is not necessary that one must have faith or religion to be virtuous, but it makes the virtuous persons a pious person if he is indoctrinated with a divine faith. All three monotheistic religions were born in narrow strip of land stretching between the Red sea and the Dead Sea; and yet a massive world population exceeding several billion in number follow these faiths. So obviously faith has played major role in the world. It has created discipline and order in many a million tribes and clans across the globe. As was aptly put by the philosopher of the East, Allama Iqbal:

Juda ho deen siyasat say to rah jati hay Changezi.

Without faith one may become a tyrant, a ruthless ruler if powerful or a beastly character if without power, living an ordinary life without any order in life.

Islam is the last religion and *Quran* the last testament bestowed on mankind. It is a book of ethics, a compendium of philosophy, logic, aesthetics, politics, sociology, law and a thousand other subjects that mankind may require in the daily pursuit. To follow it along with the Sunnah of the Prophet would guarantee happiness, joy, and satisfaction. That ultimate result may be called an abode in *Jannat* as a believer or Nirvana for some, solace for others and so forth.

Good has always been confronted by evil. For each successive generation of a reformer there has been a reactionary. The poet wrote quite rightly:

Sateezakar raha hay azal say ta imroz ... Chiraghe Mustafavi say sharare buLahabi.

Human history has documented a plethora of episodes in nearly all cultures and all time zones, where evil forces have defatted and demolished the good doers (temporarily), but as the time passed on the evil forces' reactionaries faded away and were only remembered as the despicable lot. On the other hand, while the supporters of good, i.e. reformers may have suffered annihilation and destruction at the time, but the world has not forgotten them, they are remembered indeed commemorated for their good deeds forever and ever. They have left indelible footprints on the sands of time, which function as the beacon of light for the world. Just like

Prometheus, Socrates, Jesus and many other reformers, in every nation one would find a hero to emulate and follow. The reactionaries who caused pain and death to Jesus can hardly be recalled in name or person, but the name of Jesus and his teachings live on.

Islam is religion of sacrifice. Ibrahim and Ismail are constantly remembered each year at the culmination of Hajj by all Muslims. Who was Namrood, one cannot remember but only with reference to Ibrahim AS.

The most dedicated pupil of Muhammad SA, Ali ibne Abi Talib suffered enormously at the hands of the enemies of Islam, and devoted his entire life to serve his master Prophet SA, eventually succumbing to the evil powers of Syria, which eliminated him. But he lives on through his teachings, his good deeds and his virtues. His son Hussain AS gave the sacrifice at Karbala that is unprecedented in the annals of human history. He passed away on 10th Moharram, 61 H in the vast wilderness of Naineva, not far from the valleys of Euphrates, the cradle of human civilisation. But even today the world mourns him on each Ashura. Each passing day makes him even more glorious.

As Josh wrote:

Insan ko bidar to ho leney do … Har qoum pukaray gi hamaray hein Hussain.

This essay is finally occluded with a couplet from Allama Iqbal:

Ayan sire shahadat ki agar tafseer ho jai … Musalmanun ka kaaba rozae Shabbir ho jay.

Further Reading

1. The Holy *Quran*
2. *Nahj-ul-Blagha*. www.nahjalbalagha.org
3. *Saheeh Bokhari*. www.sahi.bukhari.com
4. *Encyclopedia Britannica* (1982)
5. *Encyclopedia Americana* (1983)
6. Durant W (1966) *The story of civilisation: The age of faith*. (1950). Vol, IV. MJF Books, NY. www.ebook3000.com/cultures-and-languages
7. Syed AA (1968) *Spirit of Islam*. www.amazon.com-spiritofislam
8. Durant W (1961) *Story of philosophy*. www.gobooke.org/will-durant-storyofphilosophy, www.amazon.com
9. Gibbins E *The decline and fall of the Roman empire*
10. Russell B (1962) Essay
11. Sartre JP (1956) *On being and nothingness*. Philosophical Library, New York
12. Marx K *Das kapital*. www.amazon.co.uk/Karl-Marks/e/BoooAMPKRQ
13. Rizvi SAA (1986) *A socio-political history of Isna-Asharis in Northern India*. Ma'rifat Publishing House, Canberra
14. Maududi SA (1986) *Islam Ka Ikhlaqi Nabavi*. Islamic Publishers, Lahore
15. Macintyre A (1992) *A short history of ethics*. Routeledge and Kagan Paul, London
16. Lewine C (1989) *Cases in bioethics* – selection from Hastings Centre
17. Russell B (1991) *History of western philosophy*. Routledge, London, p 185, Reprint
18. Shiaism in the medieval period AD1000–1500. Momen M (1985) An introduction to Shii islam. Yale University Press, pp 94–95

Ethical Theories and Metaethics

2.1 Ethical Theories

Let us examine various aspects of ethics. Ethics grant us the knowledge to synthesise, analyse, defend, explain, recommend or reject a thought or create a plan of action to follow the right path and avoid the wrong path. It differs from religion and law in some ways. While religions and laws give absolute, final and definitive dictates, ethics focus on issues attempting to clarify dilemmas and puzzles, providing answers and options to resolve a situation.

The fundamental rule in ethics is the source and the validity of values. Plato believed in the absoluteness and universality of values, irrespective of times, conditions or places. For instance, truth has to be truth, whatever the situation. But many other philosophers, like Pythagoras, challenged the notion of universality of values and believed in the philosophy of ethical principles being relative, that is, dependent on such variables as time, place or customs. The theory of relativism versus absolutism continues to baffles readers even today.

Medical ethics is a form of applied ethics based upon the teachings of philosophy, humanities, religion and law. It is a practical form of theories of general ethics applied in specific situations. A recent development, more a metamorphosis of medical ethics into a new nomenclature, is currently gaining ground. It is called bioethics. It is a broader definition and encompasses a wider field dealing with health-related issues, including medical care, research, health economics, and medical management, social work related to patients, genetics epidemiology, legal medicine and several other aspects of health care.

Ethics has many components; however, the most widely accepted contemporary thought focuses its discussion under the captions and titles of 'metaethics', 'normative ethics' and 'applied ethics'. The most significant arm of ethics and indeed the most basic component is 'metaethics.' It attempts to investigate and find answers for universal truths, for example, 'the God's word, His wisdom,' His principles and norms.

Normative ethics is an extension of ethics into exploring ways and means to arrive at morally correct decisions and to regulate the standards and the parameters

S.H. Zaidi, *Ethics in Medicine*,
DOI 10.1007/978-3-319-01044-1_2, © Springer International Publishing Switzerland 2014

that determine conduct, whether good or bad. It also practically guides us into the pathways of, for example, finding truth and justice and avoiding falsehood and tyranny.

Applied ethics identifies and determines specific good and bad acts in specific issues of day-to-day life. The illustration of what we see in dealing with issues like stem cell research, assisted suicide, and managed health care are day-to-day issues covered under bioethics.

Theoretically, ethics has three branches but a single stem. The differences are subtle but visible and palpable to a serious student, though imperceptible enough to merge into each other. The final objective, of course, is to differentiate between good and evil.

Metaethics may be defined as study of the origin and meaning of the philosophy and concept of ethics. It is somewhat more intellectual and rather ambiguous when compared to the other two arms. There are numerous components of metaethics, but for our purpose we shall confine ourselves to the two basic factors, namely (a) whether morality is confined to human beings only or whether it exists in other forms of life and (b) the intellectual basis of our moral deeds and the factors that influence them.

Metaethics is also a study of universal truths. There are two directions in which philosophers guide us in the process of explaining the working and application of metaethics. One of them is the 'principle of divine command'. God exists and has given us clear mandates about good and evil. He has also granted us 'intellect' and 'reason' to choose between the two. And there is also the concept of 'rewards' and 'retribution', which shall be discussed later in the chapter about justice. This theory informs us that God is omnipotent and omnipresent. He commands and the deed is done. In Quran it is mentioned in *several places* as '*kun fa yakun*' Quran, 2:117; 3:47, 3:58.

The second concept is called 'moral relativism'. There are two forms of relativism discussed by the wise men of philosophy. One is 'individual relativism,' that is, an individual may develop, design and apply his own moral standards, depending on his individual intellect. Thus, a superhuman may cultivate high standards that may suit him but may not be achievable by ordinary mortals. The second form is called 'cultural relativism,' which is based upon the notion of 'morality' firmly ingrained in and approved by a society. It therefore implies that such a form of morality is a collective issue and is based upon such external factors as customs, traditions, faiths, beliefs and so forth.

The intellectual or psychological matters related to metaethics refer to our appreciation, analysis and motivation of factors or attitudes that guide us to be good or moral or otherwise. The psychological basis is not a new concept at all. As anyone who has indulged in debate, discussion or even study of ethical principles should know, it is a regular and routine question posed in such debates. Why do 'good' or be 'moral'? The world is essentially a bad place, and being good or moral would not bring about an iota of change in the existing ills of the society, so why even try it?

Well, the simple answer has to be based upon one's own intellect and human instinct. For instance, in response to the advice 'Do not commit a murder', the obvious response of a normal human being would be to agree and follow the advice. The consequences of such an evil act would obviously be a long stay as a state guest or, in some societies, a swift despatch to hell. Likewise, why tell a truth or not tell a lie? The obvious answer is to save yourself embarrassment or, even worse, more serious harm, that is, punishment. Why do good? Well, the answer is to enjoy a peaceful, happy and comfortable life. So the debate goes on.

Most religions have used the principles of metaethics through the ages.

One aspect of moral philosophy based upon psychological and intellectual impact is the principle of 'egoism'. Many philosophers believe that because of the inherent nature of human beings, selfishness often motivates one to do good. For instance, one may perform an act of charity based upon his notion that it is a good act, and an act of kindness, but sometimes a shadow of egoism may be influential in such a deed. Self-promotion or recognition by peers or the community of one as a good and charitable person may be the driving force, and not so much the act of charity per se. Such a form of egoistic display, even for a good cause, not only tarnishes the personality but also demolishes the whole act, at least in the eyes of religious philosophies and the philosophers.

Then there is the concept of 'joy' or pleasure in pursuing a goal through a charitable act or performing an act of kindness or benevolence. And that action to achieve happiness through a good act is definitely commendable.

Moral philosophy also grants special significance to the roles of emotions and reason. Most decisions in life are and, in fact, should be based upon logical thinking; however, some may be made on a purely emotional basis. Should emotion be the master or a 'slave'? The argument continues. Allama Iqbal, the poet philosopher of the East, wrote:

lazim hay dil key sath reahey pasbane aql
Laykin kabahi kabhi usay tnaha bhi chhor day.

Man is created in the image of God, so says the Bible, and the last divine testament clearly dictates that mankind is the *khaliffa* fil *arz*, that is, 'the representative' of 'Almighty on earth.' Mankind cannot have more glorious a status than that, and yet since he is fallible and easily succumbs to the enemy, that is, Satan, Allah has guided him throughout his life with His clear verses in the Quran.

Since man is a true representative of the Almighty, he must harness his faculties of reasoning and intellect to make decisions and not merely decide his matter based on emotions. And yet emotion is such an important and integral part of human nature that without it little difference would remain between an android and a man. So the answer may perhaps lie in control and self-restraint and employing *ilm -o- idrak* when making important decisions, as the emotion is an inferior faculty compared with intellect and reasoning.

Ali ibne Abi Talib said, "Anger begins with emotional absurdity and culminates in regret." David Hume, a renowned British philosopher, is known to have said, 'Reason is and ought to be a slave of passion'. How woefully wrong a philosopher could be!

Compassion and benevolence are two virtuous deeds. Reason would dictate that the poor be helped and the hungry be fed. So here is an example where emotion and reason have mingled together to create an act of kindness. In other words, emotions have dictated reason to do a good act.

Iqbal is famous for his beautiful poetry and thought-provoking philosophy. He granted a superior position to the attribute of emotion when he wrote

Bey khatar kood para atashe nimrod mein ishq,
Aql hey mahavae tamashe labe baam abhi.

Reason and emotion are therefore both integral parts of human nature and quite easily compatible, indeed, complimentary, in many ways.

Immanuel Kant left a lasting impression upon the subject of moral philosophy. He is one of the great philosophers who have dominated the world of thinkers since the eighteenth century.

His theories somehow appeal to common sense and are instinctive in many ways. He disagreed with the emotive theories, albeit granting minor leeway in allowing decisions to be influenced by emotions, and strongly advised that all decisions must be based upon gainful employment of intellect and reasoning. He resists all forms of emotional swings. Indeed, he believes that if emotions overtake reason, an act may not remain truly moral as the personal interest may overshadow the act of goodness.

The holy Quran has called the mankind '*Ashraful makhlooqat*'. Arabic scholars inform us that the pronoun adjective *ashraf* is a superlative degree of *shareef* and is used selectively, such as in 'Najaf- al- Ashraf.'

Aql or intellect and *Idrak* or reasoning are bestowed by Allah to mankind through Ilm, that is, knowledge, which is what the *Ibless* was kept deprived of, as is narrated in *Surat al Baqr*. *Ibless* was commanded to prostrate to Adam, indeed not the clay that he was made from, but to the Ilm that was granted to him. So the moral of the story is that *Ilm o Aql are* the ultimate attributes possessed by mankind and should be duly used.

Reason has to be the mainstay of moral philosophy. One recalls the famous story of a battle in which Ali Ibne Abi Talib overpowered the opponent in a duel but did not slay him, as the coward opponent threw his saliva on the victor's face, which in the case of ordinary mortals almost ensures a vengeful, fatal stroke on the defeated one, but Ali got up and challenged him to battle again. When questioned by the opponent, who was simply flabbergasted, Ali quietly replied, "I do not fight for myself but for Allah, and had I slain you after your disgusting action, I would have acted upon personal emotions, and that is not what I have been taught by my teacher the Prophet (PBUH)."

Emotions, however, do come into play in our actions, as indeed they should, yet should not overpower the faculty of intellect and the power of reasoning.

Another interesting debate is that on the subject of male versus female morality. Since nearly all philosophers and reformers, saints and savants, barring a few, have been male, so the moral philosophy has also been male-oriented.

Martial arts, the rules of war and the negotiation of peace treaties, property matters, even marriages have forever been male-dominated. The pattern is changing now, as more and more women have taken up important roles in society and beyond the four walls of their homes. And there are many renowned philosophers and moralists who are now at the front end of the long queue, promoting women's rights.

According to some feminist moralists, there is yet another facet to the moral theories. It is called the 'value theory'. Women know more about nature and the divine ways in matters of day-to-day life issues, such as giving birth, feeding babies, nurturing them, raising them and teaching them the fundamental rules of life. How true is the statement that the first lesson that a man learns is at the knee of his mother. Islam has granted unparalleled status to women and particularly to mothers. Allah has commanded us not to utter so brief a word as '*uf*,' that is, fie, against our parents, and we have learned from the Prophet's traditions that he granted the highest regard to the mother, far above that of the father. Accordingly, the moralists following this theory believe that women can handle natural matters and solve the moral dilemmas arising out of them more effectively than men, as they are closer to nature. There is a strong case in favour of such moralists.

Now we will examine the concept of 'normative ethics.' It involves making decisions which are moral and ethical and determining the nature of conduct, that is, good or evil.

Normative ethics is more like a road map, a navigation system for guiding one though the maze of alleys and streets in a complex world of rapidly changing scenarios. The golden rule of normative ethics is not only time tested but, in fact, as ancient as time itself. It is Biblical and, indeed, ancient in its concept and format. Do no evil, hear no evil, and see no evil is the dictum taught to every child at the early stage of his life across all cultures. Mirza Gahlib said so aptly in one of his ghazals

mat suno gar buar kahay koi,
mat kaho gar bura lagae koi....

And in another couplet he duly highlights the inherent faults that we all have:

Galib bura na man jo waiz bura kahay.
Aisa bhi koi hay jisey acah kahein sabhi

The guiding rules of normative ethics are very basic and almost instinctive in nature. What I would not like to be done unto me, I shall not do to others, and so forth.

Normative ethics is best discussed under the caption of strategies that should guide us to arrive at definitive decisions. These theories are:
1. Virtue theories
2. Duty and Obligation theories
3. Consequential theories
Although each one is developed to provide us with the ways and means to do well and avoid evil, each one has its merits and short comings.

2.1.1 Virtue Theory

Virtue theories are the oldest and most publicised theories of all. Virtue is basically a foundation stone of a good character. But what exactly is virtue?

Plato famously said, "Knowledge is virtue." It encompasses so much that no matter how lengthy a debate may be, one cannot fathom the depth of the ocean immersed in this brief statement. Plato was Plato for obvious reasons and has dominated the world of thinkers and philosophers for centuries. He was Platonic!

His student Aristotle imbibed every richness of knowledge from the master and added huge amounts of his own thoughts into what is described as Aristotelian principles.

He has defined and elaborated Virtue ethics in his famous book called *Nichomachean Ethics*. The fundamental principle underlying Virtue theory is that a person is, by nature, moral and therefore cannot commit an evil act.

Virtue is such an abstract entity that it has many dimensions and several facets, each with its own strong and weak points. It has baffled such greats as Socrates, Aristotle and Plato, as well as the modern philosophers like Macintyre, and no one seems to come up with an absolute definition of virtue. Each has defined and studied virtue according to his best perception.

For Aristotle, virtue meant good habits that control and regulate our emotions. He has given the finest details of his concept of virtue, arguably the best and the most appealing to a student of ethics, like this writer.

He clearly draws a line between a virtue and a non-virtue; through an analogy of excess versus deficiency, he arrives at the concept that virtue lies between the two extremes. So vague yet so very appealing.

For instance, rashness is an excess and cowardice a deficiency, so virtue is in between these two extremes. For him, therefore, courage is virtue and rashness as well as cowardice the lack of it.

Thus, he goes on to other attributes like temperance, liberality, magnanimity, controlled ambition, patience and forbearance, righteousness, truthfulness, etc. It is simply astonishing to read his works and simply admire his wisdom at the deep knowledge of human mind and habits that he so vividly paints.

His confusion was that he did not find it easy to clearly project man's values between the extremes in all situations, as there certainly are certain gray zones in between, such as telling a lie or withholding truth.

Plato's definition was the finest, the simplest and, to most, the best. That 'Knowledge is Virtue.' Knowledge indeed is the beacon of light that guides through darkness and saves one from falling into pitfalls and the abyss.

Plato emphasised these four traits as virtues, which are often called the Cardinal virtues:

1. Wisdom
2. Courage
3. Temperance, and
4. Justice

Of course, there are numerous other characteristics of goodness described as virtues, e.g., fortitude, generosity, self- respect, good temper, sincerity, honesty, etc.

Many theologist philosophers have these three fundamental attributes as virtue:

1. Charity
2. Hope, and
3. Faith

Alistair McIntyre is a modern philosopher. In 1981, he wrote his famous work called *After Virtue: A study in moral theory*. It is considered to be an extremely influential contribution by a philosopher in modern times.

McIntyre believes that the Greek concept of virtuosity was the by-product of the times. Since the Greeks did not have any divine source of guidance, like the Torah and the *Bible*, they were obliged to evolve their own codes of conduct, based upon the current beliefs and myths.

Hindus believe in certain myths based upon the stories narrated over millennia and described in their books such as *Mahabharata*, which provide readers deep and thoughtful insights into their beliefs and practices. Similarly, the Greeks had their own epic stories often narrated in the form of par excellence poems, like the *Iliad* and the *Odyssey*. Based upon them, the Greeks developed their theories of ethics. They were not divine and hence did not receive universal acceptance.

Greeks had small communities to handle and even smaller conflicts to cope with. Undoubtedly, there may have been major episodes in those times, like the story of Sparta, but compared with today's world, they were significantly smaller events. But that does not mean that they have not left indelible impressions on the sands of time. They certainly have. After all, events like the tragedy of Karbala, after only 40 years of departure of Prophet SA in a remote piece of *nainavah*, changed Islamic history for ever. It provided not just the world of Islam alone but entire humanity with the gold standard rules of Virtue against Vice.

McIntyre believes that Greeks contributed immensely to the development of philosophy of virtue by granting definitions through acknowledging physical strength, courage, friendship, genius and loyalty, etc., as the parameters that do not need any boundaries of faith or religion.

When the smaller hilltop communities transformed into larger metropolises and city states, and the tribal culture changed and was partially replaced by major communities, the definition of virtue also needed to be modified. The qualities desirable to possess in small hilltop communities of the ancient Greek era can still withstand the onslaught of time, albeit with some modifications. Instead of having individual physical strength, you may need ingenuity and strength of technology. However, virtues like generosity, camaraderie, shared strength and loyalty have not changed one bit over the millennia. They still remain unchallenged virtues.

In Athenian culture, courage was considered to be a major virtue; so were friendship and camaraderie, and fierce loyalty to one's tribe, people and community at large. Justice in both its forms, i.e., retributive as well as distributive, was given a very high place amongst the list of virtues. That means that the concept of 'rewards and punishment' is mighty old.

Similarly, attributes like temperance and self-control, self-denial, altruism and austerity, also played extremely important roles in the development of a personality, as well as in achieving higher goals like a disciplined, organised community worthy of mankind.

In an even larger community, cunning was not a desirable trait, but wisdom was, which was duly inculcated through discourses and open debates that the Greeks so aptly conducted.

McIntyre notes that the period of human history in the eighteenth century, so fondly called the 'enlightenment,' saw the demolition of moral conduct.

For Hume, another great Western philosopher, morality was an expression of emotion and of passion. Kant believed that moral salvation rested with human rationality. However, Bentham and Mills' utilitarianism, along with Kierkegaard's existentialism and Kant's deontology, have all but failed to convince McIntyre.

He said that Greek Greats were the genuine founding fathers of Western thought, and since they had no divine books or prophets to guide them, they had to cultivate their own moral principles based upon local folklore and mythology. Therefore, the question that comes to mind is, "Do we need the Greek philosophy to guide us?"

No, as Muslims, we have the Quran as the ultimate divine book to guide us, and the Prophet's Sunnah to show the practical ways of applying those guiding principles in all matters of life and death. Since we practice medicine in the West or at least employ Western methods of treatment irrespective of our habitat, we have to learn the Western principles of ethics, and they originate from the Greek thought.

An interesting objection raised by Richard Bernstein in his *Philosophical Profiles* was that McIntyre placed undue faith in the works of ancient Greeks and rather hastily discarded the later-days philosophers like Kant, Bentham and Mills. Perhaps the truth may be found in between the works of ancient Greeks and the modernists.

2.1.2 Duty Theories

The mother word *Deontology* is derived from the Greek word *Deon*, or duty. We know that there are certain obligations for each one of us to perform as individuals, and then there are some duties that we must perform as a community. So what are these duties that we should perform in our daily lives? And how do you determine our 'goodness of actions'? Theories that explain all or most of these phenomenons are called 'NonConsequestionalist theories.'

Out of these, perhaps the most relevant to medical ethics is Kant's theory of Deontology.

Kant believed that as human beings we must be rational in our deeds and actions. We must grant due priority to reason as against passion; and as human beings we are obliged to do certain duties to ourselves individually as well as collectively. Kant firmly believed that mankind is supreme amongst all other creatures and duty-bound to do good above and beyond personal desires.

Kant believed that an action must be based upon good intentions and good results. In other words, the means and ends should be the same, with both meant to achieve goodness and welfare of all creatures big or small.

Human dignity, self-respect, honour and autonomy are the key points of Kant's philosophy of ethics.

Then there are the 'Consequestionalist theories.'

These theories determine our moral responsibility by judging the outcome of our actions. Sometimes they are also called 'Telogical theories.'

And they can be divided into following categories:

Ethical Egoism

It means that an action is right if the outcomes/consequences are more favourable than otherwise, only to the person performing the act.

Altruism

It is best translated as '*Esaar*,' and it means that the consequences are more favourable to everyone else except the person performing the act. It is best understood as the concept of self-sacrifice for the pleasure and welfare of others. It is an extremely common practice in Eastern culture, where parents are constancy depriving themselves so that their children can benefit with the goodness of life, acquire higher education, better skills, better quality of life, etc. Ask any vendor on the road in Indo-Pak, his first priority would be to provide better life, home and education to his offspring.

Utilitarianism

It measures the significance and importance of an action by measuring its consequences. According to this concept, the good should be a general one meant to benefit the society, not just an individual, and bring pleasure to the largest possible number of people. It means that the consequences are more favourable for and less harmful to a larger population. It also means that in order to achieve a certain goodly result, one may employ 'less than good' means. Alternatively speaking, it also implies that the 'the end can justify the means.' For instance, in order to save the community from evil, one may put a thief behind bars so that the community is saved from his evil acts. On the face of it, it sounds very appealing, but there are many experts who thoroughly condemn the philosophy of utilitarianism, since often enough it can be abused. Imam Ali, condemned a good that came out of an evil act!

Bentham was the first proponent of this theory and he described it under two subheadings:

A. Act Utilitarianism, which implies that the consequences of each action may be tallied to determine whether or not an action is morally right. It also means that the end is more important than the means.
B. Hedonistic utilitarianism, which means that the pleasures and pains are the only consequences that matter in determining the moral status of an act. Therefore, if the final outcome of an action is pain, then the act is unethical, and if the outcome is joy, pleasure or happiness, then the act is moral and ethical.

Theories of ethics continue to dilate, as we immerse into the ocean of knowledge. I must, however, stop here, as my essay does not give me the luxury of undue expansion. A serious reader is therefore referred to philosophical treatises and books on ethics, which are plentiful and now freely available on the internet, even on a mobile hand-held device, mostly free of charge. Instead of people travelling long distances to the libraries, the literature and the knowledge base has come to the reader. Such are the wonders of modern technology. It would be unfortunate if one did not avail oneself of these opportunities to gain knowledge and continue to gain more knowledge throughout life.

2.2   Metaphysics and Metaethics

2.2.1   A Secular and Religious Perspective

Logic and metaphysics are two essential components of philosophy. Logic is one scientific way of debating and proving one's point of view in a discussion. Metaphysics is more delicate, subtle and less argumentative and quite pleasing to the soul. It provides a true sense of solace, comfort and peace to one, without the hectic approach of logic.

Metaphysics is of particular relevance in such debates as morality and religion or, indeed, the lack of it. It is a way of scientific study of the proven as indeed, supernatural, paranatural, hypothetical thoughts, ideas, principles and norms. For instance, the whole debate on soul, its existence, its entry into human embryo, its weight, mass if any, and its departure and final abode after death are some interesting debates that have puzzled even the likes of Aristotle, let alone lesser mortals over millennia. Aristotle introduced metaphysics as way of looking at and debating the issues relevant to physics. In other words, metaphysics means 'after physics.' So it is basically a way of hypothesising or reflecting on the issues dealt by physics.

Metaphysics is a science which was primarily introduced by Aristotle, that grandfather of Western philosophy who seems to have his fingers in all pies. This science includes the study of many subjects; the salient amongst them is, first, Ontology, i.e., the discussion revolving around the concept of existence and being. Many philosophers over the ages have debated on the subject and enlightened novices like me, although no definite conclusion could be drawn. Jean Paul Sartre wrote a book dedicated to the subject called *Being and Nothingness*. I read it some years ago and found it extremely hard to swallow, mind boggling and leading one to no definite conclusion. The sum total of it all was that nothingness is what seems to work like the black hole in the space, that seems to be a reality; all the rest is a non-reality, even our day-to-day living, eating, sleeping, reproducing, etc. The change in our existence baffled many scientists and philosophers, including the likes of Darwin and his diehard opponents presenting the famous theory of evolution or the lack of it. The big bang theory is once again gaining fresh system since the momentary appearance of the so-called God particle. So the debate goes on.

The concept of *Jabro ikhtiar* or freedom versus compulsion is another issue. It has been discussed elsewhere in this book.

The second principle of Aristotelean metaphysics is called the 'Natural theory'. It revolves around the time-honoured discussion on the existence of God or a supernatural deity. The philosophers of earliest years, including Socrates and Plato as well as Aristotle, believed in the existence of a supernatural deity, though they also supported the Greek gods, so perhaps it was a cultural rather than a faith-based practice. Later, most of the Western philosophers, who were practicing Jews, defended the existence of one God as taught in the holy Talmud and the teachings of Moses, who communicated with God in person. Philosophers like Bergson, Spinoza, Scheaupenhaur, and even earlier philosophers not necessarily Jewish, like Thomas Acquinos and Bacon, all seem to have faith in God. It is, however, the authoritarian and dictatorial approach of the Church, the Popes and the Cardinals, who monopolised the collection and dissemination of knowledge, ordering execution or putting the writers on stakes to be burnt alive, etc., a rebellious attitude developed in the post-Renaissance period. Some of them became totally anti-God but continued to benefit mankind in their own way, which today is not just good for mankind but indeed much better than some of the original writings of the Mullah or the Rabbi or the Priest. It once agin proves the point that morality is not the sole property nor attribute of a religious person but is above and beyond the religious faith of any kind.

One fine example of a modern philosopher is Richard Dawkins. He is totally convincing in his arguments. He wrote a book called *The God Delusion*, calling nearly every great philosopher either Godless, anti-God or an atheist, a non-believer. The list includes anyone from Galileo, Darwin and Einstein to the contemporary thinkers.

Imam Ali was asked if he saw Allah or did he pray to Allah in blind faith, to which he replied that he did not worship a deity unless he saw it. He went on to say that His visual signs are available in all creatures small and large. Ali gave a sermon on the creation of the universe, which is yet another masterpiece of Ali AS, included in *Nahjal Balagha* as Sermon 1. The reader must read this sermon to obtain comprehensive and descriptive information on step-by-step evolution of earth, stars, moons, the winds, the oceans, the vegetation, and the whole universe.

The third element of Aristotelian metaphysics is called the principle of universal sciences, i.e., logic and discussion on the opposites and contradictions. This particular field is of immense value and significance in the *Ijtehadi* schools called *Hwaziahe Ilmia* of the Shia faith. Logic is not easy. It demands comprehensive knowledge of the subject before one can discuss it. Any argument that is revolutionary and formative begins with negation.

The best example that can be given to support this theory of negation is seen in Islam. The beginning of Islam is 'La,' i.e., No or none. *La Illaha*. It means there is No god, but Allah ....., or *La Raib fil kitab...* . There is no falsehood in this book, i.e., the noble Quran, or indeed *La haul wa la quwate...* i.e., there is no power except that of Allah and so on.

Another important field included in metaphysics is called Cosmology. It is a subject that is mind-boggling, fascinating and, for a believer, provides ample evidence to further augmenting his belief in the only creator of such marvels. Dr. Brian Cox shows such amazing pictures of space and gives such wonderful scientific explanations that one cannot leave the TV when he is on the tube. The universe is expanding each moment, and the process of evolution goes on unabated. The black mass and black hole have baffled even the masterly minds of no less a scientist than the living legend Professor Stephen Hawkins. And what's more, the spinning action in the atom and its minutest fractions like neutron, proton electron and bosons, confirms the fact that life is present in each and everything in the universe or space. Either it's a vegetative or a non-vegetative life, but the spin of an atom just shows that there is life everywhere. But the human mind and technology are still in their ice age, and much snow has to melt down before humans can achieve what they have always endeavoured for.

Recently a friend, who is an oracle, a mountain of wisdom, published a book covering many subjects, including metaphysics. He honoured me by inviting me to write its introduction. Since it is relevant to the discussion in this chapter I am taking the liberty of reproducing it. His book, called *Nairange Haqeeqat*, is an excellent example of scholarship. Some of its chapters deal with issues discussed in my essays; hence, it may be worth my while to include the abstracts from the book in the form of my introductory note.

'I met Syed Sahib in the midlands in 2006. It was either a coincidence that we shared the same name or perhaps both of us are part of that minuscule particle called 'photon,' which remain entangled as a pair, though each one could be a part, residing anywhere else in the huge universe. He has discussed this fascinating topic called 'Entanglement Theory' in this book. Suffice it to say that we developed mutual affection, which has gradually progressed over these past few invaluable years. Every time I meet him, I learn something new and feel humbled on my deficiencies and inadequacies in *Ilm- o - idrak*.

Syed Shabeeh -Ul -Hasan is a man of all seasons. He has banking and finance background and enjoys his well-earned retirement in the pursuit of *ilm- o Irfan*. He has taken up the task of spreading knowledge through his writings. He has written extensively upon many diverse subjects. They are regularly published in periodicals in the UK and elsewhere. He has total command on his mother tongue Urdu, which was nurtured by such greats as Vali Duccani, Meer o Ghalib, Anis o Dabeer, and later on flourished under the tutelage of the rulers of Delhi and Luknow. In his own words, the credit must be accorded to the Punjab for keeping the language thriving, indeed, flourishing, for which thanks are due to Iqbal, o Faiz o Faraz.

Urdu, a true representative of the culture and civilisation of the valleys of Gang-o Jaman, is his medium of expression.

After the first reading of Shabeeh Bhai's essays, I was obliged to warn him in the words of Mir Anis:

Anis dam Ka bharosa nahin thaher jao,
Chiragh lay Kay Kahan samney Hawa Kay chaley.

Due to the possibility of facing adversity and gravity, Allama Iqbal had consoled his symbolic '*uqab*', as follows:

tundiay baade mukhalif say na ghabra ay uqab:
yeh to chalti hay tujhay uncha uranay kay liay.

Shabih Bhai's *uqab* actually looks for the '*bade mukhalif*' rather than *avoids* it. And his *uqab* does fly higher and higher as he unfolds one story after another in this book.

Some of his stories are bound to disturb if not outright upset the odd, myopic reader. But artists, thinkers and story tellers do not paint or write for the pleasure of the masses. Not his kind, anyway. They write because they carry a sensitive soul and an even more sensitive mind. They notice a wrong thing around themselves and feel compelled by their inner self to document it. That is the way they lodge their protest. They don't care if the Sultan would approve and applaud or lose temper and punish. They simply air their feelings, as if to lighten the burden on their conscience. Such people are like a candle burning at both ends. They suffer at the hands of their intellect. These folks are forceful, effective, passionate and visionary. They leave their indelible footprints on the sands of time. Syed Shabeeh- ul Hasan is one such writer.

He has divided his essays into several chapters. He has written on the extremely sensitive subject of creation of the universe and its creator. Exhaustive research is carried out into the evolution of mankind from its initial stages from the primordial cell to the modern human being over several million years. The physiological and chemical composition of life per se is discussed in many essays. The role of RNA and DNA in the development of a person is masterfully discussed.

On a recent visit to the Think Tank Science museum in Birmingham with my grandsons, after reading his essays, I looked at the DNA model on display, with some intent, and at its various components like Carbon, Dioxiriboneucleic acid, and Phosphorus. I am amazed that a man with the background of finance could so authoritatively discuss such delicate, minute and intriguing details of DNA, the chromosomes, the mitochondria, etc., and their role in biological development of man and mouse.

Richard Dawkins is a contemporary philosopher of international repute and recently, he Published a book called, 'The greatest show on earth'. One particular essay that impressed me most was on the life cycle of the Carbon molecule. Briefly, it says that the Carbon atom, which was created at the beginning of the universe, is still alive and kicking. It changes its form from a building block in a living cell to the charcoal or a diamond in the bosom of the earth, or from the glaciers sleeping on top of the majestic Himalayas to the Chlorophyll that feeds a plant, to almost everything else in the world.

Syed shabeeh has described the human life cycle in his unique style, which is significantly superior to the article quoted above. He has gone in depth into chemistry, then delved into philosophy and finally related it to the subject which is closest to his heart, i.e., Cosmology. He quotes a quartet from Rumi, which was written many hundreds of years before the article on the Carbon life cycle mentioned above, and several centuries before Newton. This Poetic stanza sums it all up, that life begins and goes through a cycle to eventually pause at one stage only to acquire a different form, and the journey continues ad infinitum.

Jalal ul- din Rumi wrote

I died as an element,
And changed into a plant,
Died as a plant and rose up as an animal,
When died as an animal, I rose up as human,
So why be afraid of death cos for, when has death diminished my existence,
I'll die again as a human
To rise as an angel,
The angel's spirit will sacrifice itself over that existence which is eternal,
And I will merge into it,
An entity that is outside the domains of thought and reason,
I declare this with the musical tones and overtones,
That my existence is not meant for non-existence,
And we shall revert to where we came from!

Thus, death is a bridge between life in this world and hereafter. The death is confined to the body but not soul, which returns to the creator albeit in *alam- e Barzakh* until doomsday.

It is the subject of cosmology which is very dear to him. Brian Cox is currently the finest TV scientist, who narrates the intricacies of the universe in a professional manner, which is simple and easy to understand. But you just have to read Shabeeh Bhai's articles on theoretical physics and Quantum theories, in which he describes the minutest fractions of Atom and its impact upon our current and future lives. His description of the invisible fractions of Atom is highly educational, masterly, mystical and magical, to say the least.

He loves outer space, the universe, the moons and the stars and the brightly shining sun and thousands upon thousands of galaxies. He points out that the universe is still growing and stretching apart, disappearing into the vast wilderness of space. Dark matter and dark energy occupy most of the universe, and only a small portion of it contains all those stars and the Milky Way, which anyone from Galileo to Stephen Hawkins to Patrick Moore and Brian Cox can see and study.

It appears that Allama Iqbal may have expressed the same fact in a couplet.

yeh kainat abhi natamam hay shayid,
Kay Aarhi hay damadam sadae kun fayakun

He wrote in an essay that for many years, the scientists have been engaged in exploring the ultimate particle, called the God particle. Only in June this year did the Large Hadron Collider buried under the land mass of Switzerland and France finally discover said particle and name it after its theoretical fathers, Peter Higgs and Surrendar Nath Bos as 'Higgs Boson.'

Shabeeh Bhai writes that now that this particle is discovered, man may be able to know how he and his universe actually came into being.

The origin of religion and its impact on mankind is another subject that he has discussed in his essays. He observes that religion has served mankind over millions of years. One may remain satisfied with his faith and belief and not bother to raise a question. On the other hand, he may explore all possible and many impossible avenues and find an answer to the time-honoured question of the existence of God and the creation of the universe. He believes that religion demands total faith without any leverage for the application of reasoning. This debate on the role of *Aql* o *Aqeeda* is an ancient debate and despite huge discussions, to borrow a phrase from Samuel Becket, is still 'waiting for Godot.' One must read his exposé on the origin of religion and the role of faith with the background of scientific explorations to understand the subject.

Perception and reality are two different elements and may be illusive in many ways. After meeting Shabeeh Bhai and reading his articles, one may perceive that one may be dealing with a non-believer, an agnostic, or even an atheist. But suffice it to say that the reality could be just the opposite; in many ways, his thoughts are echoed in the poetry of Ghalib, whom he simply adores. The similarity between Shabeeh Bhai's thoughts and Ghalib's poetry makes me feel blessed to have read them both.

Like any other lover of Ghalib, I too have regrets that I missed out his era. But dare I say that remorse is diminished on reading Syed Shabih -ul Hasan's essays, because if I were born then, I would have missed Syed Sahib's writings, whereas I can still enjoy Ghalib's prose and poetry through his documented works.

Ghalib was also an anti-establishment person, a thinker and a philosopher who saw nature very closely, beyond the vision of an ordinary folk. Thus, he warned us

Hain kawakib kuch nazar aatay hein kuch,
Daiaty hein dhoka yeh bazigar khula

Ghalib knew in his heart, just as Shabeeh Bhai does, that existence of God is not an issue. He exists, albeit we just try to reach Him by according Him shapes or forms. And that is wrong. In fact, I can clearly relate to Shabeeh Bhai's faith through Ghalib. Both are believers but have strong objection to anthropomorphism. Shabeeh Bhai quotes that Allah is closer to one's Jugular vein, as mentioned in Quran. In other words one does not have to go looking for God on His arsh as Ghalib opined:

manzar aik balandi par aur hum bana saktay
Arsh say paray hota kash Kay Makan apna

And further he wrote:

hay pray Sarhad e idrak say apna masjood
Qiblay ko ahle nazar qibla numa kahtay hein

And then he wrote:

Emaan mujhe rokay hay jo khenchey hay mujhe kufr
Kaaba meray peechay hay kaleesa meray agey

Further on, to display his humility and the omnipresence of Allah, Ghalib wrote:

Na tha kuch to Khuda tha kuch na hota to Khuda hota,
Duboya mujh ko honey nay na hota Marin to Kia hota

Ghalib, as we know, was a believer but, because of his lifestyle, felt guilty on going to the mosque, as he said:

Kaabay kis munh sey jao gay Ghalib,
Sharm tum ko Magar naheen aati.

It is not easy for anyone to write on such delicate and sensitive topics unless you have a thorough knowledge of the subject, and I dare say Shabeeh Sahib merits to be labelled as an authority in more than one way. It is said that you must read a hundred pages to write one page.

As mentioned above, the existence of God has been a matter of debate ever since mankind acquired sense and sensibility. Syed Sahib has devoted a fair portion of his book to this ultra-sensitive subject. I leave it to the reader to make up his own mind about the dilemma after reading his essays. One thing that I must say, however, is that with his powerful writing he can surely shake up anyone's belief!

Shabeeh Sahib has deliberated upon the role of many a philosopher of eminent stature, such as Nietzsche and Goethe, on the human mind. He has pointed out that Allama Iqbal, during his stay in Germany, was deeply influenced by these German philosophers, thus harnessing the concept of a superman whom Iqbal called the *marde momin*, after 'converting him to Islam', and the concept of self esteem or *Khudi*. Syed Sahib quotes,

Khudi ko kar baland itan kay, har taqdeer say pahlay
Khuda bandy say khud poochay bata teri riza Kia hay.

It is an interesting observation and historically correct.

I believe Aristotle was the first-ever philosopher to search for a 'perfect human being'. He has defined eleven attributes that his 'ideal man' would possess. Regrettably, he failed in his pursuit, as did many others. Did Iqbal find one? Maybe. He wrote, while reminiscing over a famous episode of early Islamic history,

kafir ho to talwar pay karta hay bharosa
Momin ho to bay taigh bhi larta hay sipahi.

Syed Sahib has eminently discussed the role of ethics and its relationship with culture. Ethicists have always debated the issue of universalism versus relativism. Many of them believe that all ethical norms like wisdom, courage, valour, truth and justice are universal attributes and applicable to the societies, irrespective of the religio-cultural variations. But many others believe that although the ethical norms are universal and no one can deny them, the impact of religion and culture cannot be ignored either, particularly in application of these principles. It is not necessary to be a religious person or a believer to be moral, nor is one bound to be amoral if a non-believer. Syed Shabeeh has discussed the significance of ethics in the life of mankind in a professional style.

In the chapter called *Jabr* -o- *Ikhtiar*, he simply excels in the philosophical and metaphysical analysis of autonomy and compulsion. He has evaluated this time-honoured debate on the omnipotence of God and His approval of granting limited power to His chosen creation, i.e., mankind. It is reported that a peasant asked the same question of Ali Ibne Abi Talib. He replied by commanding the peasant to lift his one leg from the ground, which he did. Then the Imam commanded the peasant to lift his other foot also. The man obviously felt perplexed at this command and asked how was that possible. Ali then said that it was the simplest way to explain to the ignorant person, that man possesses *ikhtiar* in some matters but not in all.

The philosophy of *wahdat* -al *wajood* has baffled a thousand scholars over centuries. Who can ever forget the slogan of *Anal -Haq* (*I am truth or I am God*), raised by Mansoor Al Hallaj, and many others who were either executed or banished for their belief? Syed Sahib has written a beautiful essay on this very subject with intensity, passion, emotion, earnestness and total commitment. I think it may be called the very *soul* of his essays.

Annemarrie Schimmel, in her famous book called *The Mystical Dimensions of Islam* (publisher, University of North Carolina) writes extensively on both subjects, namely the philosophy of *wahdat al wajood* (P 267–269) and AlHallaj. A serious reader is best advised to refer to this masterly exposé by a lady who was German and wrote extensively on Eastern subjects. She also gave the best possible translation of the word '*Umi*' with reference to Prophet as 'untaught,' unlike all the rest who have translated as an illiterate. It is highly degrading to the prophet to even think of him in those terms. He was all knowledge, reason, ethics, morality and religion, combined in one personality.

Syed Sahib has also discussed the eternal subject of *Rooh* at some length which is simply too informative to be left unread. It is true that physicians do not know when and how the *Rooh* enters the human body. In olden days the Pineal Body, a tiny human organ sitting at the bottom of the brain, was considered to contain the soul. But we know it is untrue. Quran defines it as an '*Amre Rabbi*'. According to a Prophet's tradition, the *Rooh* enters the embryo after 120 days of conception. Shabeeh Sahib has discussed the process of fertilisation and ensoulment, right from the phenomenon of the solitary sperm out of millions succeeding in fertilising an ovum, raising the question of Justice. He poses the question, "What could be the criteria for the selection of the single sperm out of millions, and why should others perish?" He asks if it is just on the part of the Creator to destroy all those millions but one.

Obviously, it is a great philosophical debate. Philosophers from Aristotle to Spinoza and Voltaire and, later on, modern philosophers like Sartre and Singer have all deliberated on these subjects, namely life and its sanctity. Modern bioethicists like Beauchamp and Childress, as well as McIntyre, the founder of the Neo- Virtue theories, have also deliberated on these sensitive subjects. Shabeeh Sahib has certainly looked at it through yet another angle, thus adding a third dimension. It genuinely reflects on the depth of his knowledge and reflective thinking.

Syed Sahib has quoted Einstein and his theory of relativity at many places. An interesting debate focuses on the subject of Time. He has given a new direction to the ancient philosophy of relationship between Time and Space. He says that time is in perpetual motion; at this very moment when I finish writing this letter, time has already moved on, and my last word has disappeared into the abyss of past. So there is nothing like present as it is all relative in its own right. Einstein is quoted to have introduced the concept of the 'worm hole' and the possibility of a future time-travel machine. It reminded me of the famous film called *Back to the Future*, which I saw in my younger days. If only one could travel into the past and look at all those historical episodes that we have read in the history books and heard in grandma's tales.

Shabeeh Bhai envisages its possibility, as he believes that human beings may actually witness the transportation of human bodies across the universe, belying gravity and the relativity of time and space in future, just as it happened in the popular TV series, in which Scotty energised Captain Kirk!

As I was reading this essay, a thought crossed my mind; was *Mairaj* an episode of similar nature, I wonder?

What impresses me most about Mr. Hasan's essays is the depth of knowledge and his insight into the intricacies of the universe and its marvels. Besides, he is so earnest in sharing the information with the readers that he makes many extremely complex subjects quite simple to absorb and comprehend. He presents them in a masterly fashion in a lovely prose supplemented with poetry.

Translating scientific discoveries into Urdu is not easy, but Shabeeh Sahib has done not only that but accomplished much more. He has traced the origin of species through a process of evolution, from the theory of big bang to the origin of first primitive life, the single cell, and so on until the development of homosapiens. He has narrated the stories of discovery of the microscope, then Ehrlich's first pathological discoveries, identification of bacteria, Pasteur's contribution to the sciences, and their role in causation of diseases, culminating with the latest discovery, namely the Higgs Boson. Shabeeh Bhai has appropriately quoted *Abdul Hameed Adm*, as follows:

takhleeq e kainat Kay dilchasp jurm par
Hansta to hoga aap Bhai yazdan kabhi kabhi,

This couplet displays the reality of his belief. In fact, Bergson has written a whole chapter on this topic, defining the fact that all biological creatures are bestowed with the instincts of security, curiosity, hunger, companionship and quest for supremacy and progress. He says that all creatures are similar except for mankind, who is given the power of reasoning. A man can therefore employ his intellect and not base his decisions on emotions or passions alone. Such men are often called the wise men.

Dr. Iqbal penned it differently when he wrote:

lazim hay dil Kay saath rahey pasbane aql
Lekin kabhi kabhi usay tanha bhi chor dey

Writing on the interesting puzzle of dreams, a subject that brought immortality to Sigmund Freud, Shabeeh sahib has described the physiology of sleep precisely and scientifically and talked extensively on the theories of dream, concluding his essay on the couplet by Ghalib:

tha khwab mien khyal ko tujh say moamala,
Jab aankh khul gai na zian tha na sood tha

If I have to choose one article from the galaxy of his writings on the philosophical topics, I would choose the one called *Mabaad altabiat 'Maghrebi falsafey Ki roshni mein'*. Entities, colour, fragrance, music, psychology, ethics, dreams, thought and imperceptible sensual elements have all their origins in one's mind. Many great philosophers like Pythagoras, Socrates, Plato and Aristotle, gave due importance to metaphysics in their times. They wrote or spoke on subjects like supernaturality, religion, concept of God, divine instructions, thought and feeling, intellect and perception, fact and fiction, memory and its various branches, and many an abstract thing that we see in our life. They discussed such topics as the wetness of water, which is a fact but also an abstract thing, or the light which can be *noori* as indeed *nari*. What is time or mass or matter or the universe? Such are the abstract but intensely essential components that metaphysics handles.

Socrates was different from all other philosophers of the ancient times. He actually refused to acknowledge the gods of Athens and claimed that he received guidance from his own God, which guides him on the right path and prevents him from an evil way. Socrates was most influential in promoting the study of such metaphysical entities as faith religion, *ruh*, truth, honesty, integrity and, above all, justice. All these elements are tangible but also abstract, and that is how the debates go on.

It is an extremely difficult subject that ordinary mortals like me would happily declare 'out of course,' even in our leisurely reading.

Metaphysics, ethics, logic, aesthetics, and politics are all offsprings of the mother philosophy. Syed Shabeeh has discussed them in the light of the writings of the Sophists like Pythagoras and Greek greats like Socrates, Aristotle and Plato, in a masterly fashion. I am a student of ethics and have had the joy of reading *Nicamachean Ethics of Aristotle* and *The Republic*, by Plato, and would urge everyone to read them. Alternatively, one may read what Shabeeh Bhai has written and get the real flavour of philosophy.

Another point that Shabeeh Bhai has made and, dare I say, not many people know, is that Socrates, even though the greatest philosopher of the West, did not write a word. Indeed, he called it an insult to human mind to write down rather than commit to memory. How strange indeed! Today's educationists totally disagree with this principle, as memory and recall are left to the machines. The mind is employed only for problem solving. Could one dare say that even Socrates could be wrong sometimes?

2.2.2 'Metaethics'

An important branch of ethics is metaethics. While ethical principles and doctrines are theoretical and rather rigid, as will be discussed later, metaethics is not concerned with advising, guiding or prescribing a road map for human behaviour to make it more ethical and moral. It is only interested in the synthesis and analysis of the diction, words, meanings and communication employed in discussing moral issues. One essential component of metaethics is emotivism, which is theory of metaethics based upon the language employed in expressing or debating moral issues and not at all on the character or the actions of a person. It basically involves a critical evaluation of an ethical discussion after it is made.

In Greek, the word meta means after, so just like metaphysics, i.e., after or beyond physics, metaethics is also a process involving the debate after or beyond certain discussions on ethics or morality.

Translating the principle of metaethics in Medical ethics, let me discuss the issue of Surrogacy. One particular physician following a religious school of thought totally condemns it. The second physician follows another theological principle and believes that surrogacy through a third-gamete transfer brings joy to the family and is permissible. Another physician, who is secular but ethical, sees no problem with the technology and strongly recommends and practices it too.

Since the matter is complex because it may yield unhappy consequences, the matter is referred to the ethics committee. Since this particular ethics committee did not have a philosopher on board, and it was a moral slippery slope that was under debate, a special request was made to a secular philosopher and also religious scholars of Shia and Sunni sect in Islam to advise. As expected, no consensus was reached. As the saying goes, wise men seldom agree and fools hardly ever disagree. The matter was put on the back burner. The *maulanas* fought with each other aggressively and literally walked out, and the secular philosopher dilated upon the deontology reasons for one physician favouring surrogacy and supported the action based upon the principle of utilitarianism described in certain Islamic circles as *maslaha*, but decided to refrain from giving a clear verdict in favour of surrogacy and used the metaethical approach of defining various theories to the physicians involved in decision making and helped them understand the nuances of normative ethics before taking leave of the ethics committee.

The physicians were engaged in what is described as the first order activity, and the philosopher was invited to clarify the terms and terminology and function as a second-order clarifier. The matter remained unsolved, but huge debates and discussions took place in order to reach a conclusion.

And that is where the role of *Marjah* comes in. In the Shia faith, a *Marjahe Taqleed is* a pious man who has a panel of experts advising and defining the intricacies of a given procedure and technology just like the philosopher in the hypothetical scenario mentioned above, but unlike the ethics committee, the Marjahe Taqleed will give a mandate or issue *fatwa*, an edict, condoning or condemning a procedure or method, in the light of Quran and Sunnah as indeed his Ijtehad. Sometimes, though, even the *Marajaeh* may disagree as is the case in the matter of third-party gamete transfer through Sigheh, discussed elsewhere in this book. In such situations it is best advised to err on the side of caution or have further directions from the Mracjaeh.

The Noble Quran is the ultimate book bestowed on mankind by Almighty Allah for its welfare. The holy Prophet Mohammad (PBUH) is the final messenger of Allah to guide us through the maze of life. He said, "I have been sent to teach you *Ahklaq* (Ethics)." Therefore, one attribute of Prophet SA is the *Mukaram* ul *Akhalaq*.

The Noble Quran and the Sunnah provide us with the guidelines that clearly dictate the terms of ethics as indeed morality. They identify the principles of *Khair* and clearly dedifferentiate them from the misguiding practices of *Shar*. In one simple phrase the Noble Quran dictates us to Enjoin Good and Forbid Evil (*amr* Bil *Maroof* wa *Nahi an-al munkar*).

Since we are taught medicine, which was developed in the West, and practice what in the layman's terminology is called Allopathic medicine, we have to know the Western thought on matters of medical ethics. However, depending upon the patient's ethnic, religious and cultural beliefs, we must also take into considerations such values and norms as is applicable to them, in their management. It is the one major reason that the philosophy of medical ethics is particularly challenging if a physician is dealing with a diversity of cultures and faiths. Practicing medicine in a pluralistic society is certainly more culture-sensitive than in a uni-faith, uni-culture nation. Adaptability and adjustment to the needs of multi-faith or without-faith societies is an art that a physician must harness. Medicine has no borders and no restrictions. It does not differentiate between human beings based upon their colour or creed. The fundamental rule of medical care is the respect to a human being. No other variable must influence the care or the treatment provided to a patient, or else the founding brick of professional ethics will lose its place and the whole facade and the structure will simply collapse. Yes, there are amoral and even immoral physicians in this world, who do all kinds of abominable acts and still wear a white coat, symbolic of purity and integrity. But they are few and far between, and sooner or later they are not only exposed but discarded from society.

I had a personal experience with such a character. Many decades ago, I had a private hospital, in which I encouraged a multitude of practitioners to treat their patients. This man presented to my administrative office a fine testimonial of a London-based institution with a degree in Cardiology. We greeted him with open arms, and he impressed everyone with his masterly knowledge and skills on ECGs. One day, however, one resident noted something odd about his clinical approach to a patient with congestive cardiac failure. It was flagged, and a watch was put on his other activities. Soon his game was up, and it was discovered that he was an ECG technician in a hospital in London and pretended to be a cardiologist with membership to us. When confronted, he simply vanished from the scene, surely to degrade himself elsewhere.

The religious knowledge is often enough sketchy in most physicians, but at least the Muslim physician has distinct advantage of the fundamental of goodness and such virtues, through the *namaz* that he must offer five times a day. Besides the noble Quran is an everlasting document available for ready reference in matters of *Akhlaqiat*. Moreover, the teachings of the holy Prophet, Sunnah and his sayings are a constant reminder to us for doing what is good and forbidding us from evil.

Western thought is partly religious, based upon the teachings of the divine books, i.e., the Old and the New Testaments, but mostly secular since the religion unfortunately lost its grip on the West many centuries ago. Therefore, most of Western thought was developed by non-believers, agnostics or outright atheists. That, however, does not belittle the significance of the work performed by many a great philosopher. Immanuel Kant (1724–1804) was one such thinker who has contributed immensely to the philosophy of ethics.

Kant was more concerned about establishing a fundamental foundation for morality than an ethical system good enough to face the social challenges. He believed that a rational mind would logically result in making decisions which are bound to be morally right, irrespective of the outcome or the consequences of such an action. He firmly believed that a good action has to end up in a good result. Therefore, the 'means, as well as the end,' both have to be good.

And how do you define 'good'? It is a debate that has troubled philosophers over generations. A 'good will' is perhaps the only thing that can be called good, at least arbitrarily. However, an intention, a will, a desire, a thought, a notion are all intangible entities. The tangible form or an action verb would be a good conduct, an actual evidence of someone's honest, truthful and practical action to do well. A display of such good attributes as truth, honesty, courage, bravery, generosity, loyalty and perseverance in day-to-day life would be a definite proof of one's good intentions and not just believing in them without their practical application.

In the field of medical ethics, the principles of good versus evil are seen in more ways than in any other field of sciences or art.

Regrettably, medical ethics did not receive its due recognition until very recently, and it is only now that it is beginning to make some inroads in the field of practical medicine.

The underlying principles of medical ethics are based upon the following theories of morality:
1. Divine command theory
2. Natural law ethics, also called the telogical principle
3. Utilitarianism
4. Deontology

While most secular scientists may separate science from the religious teachings and medical care from the divine dictates, such as the command to serve the sick, the monotheistic religions firmly believe and command their followers to incorporate the moral code in the religious doctrine.

In the West, where religion has all but vanished and Christmas is no more than a time of festivity, the East continues to maintain close bonds with the religious teaching. Hence, the Western concept of medical ethics is so different from the Islamic philosophy of medical ethics. That does not, however, mean that one can simply discard the Western thought, because the medical practice that we are taught and trained to practice is primarily based upon the Western principles of medical education. Time alone would clearly demarcate the principles of medical education that the East must adopt to cater to its needs and within the parameters of religious thought. Until then, anyone practicing medicine has to learn the principles of medical ethics from the existing Western thought, albeit modify them depending upon the individualistic cultural demands.

Amongst the Western thought on medical ethics, Kant's philosophy of deontology appears to come closest to the Islamic teachings.

Deontology regards duty—doing what is right, because it is the right thing to do, as the fundamental block of morality.

Kant believed that a person should perform his duty purely of the good will irrespective of the lure of the rewards that such an action may bring forth or the fear of punishment that might ensue. He also believed that Reason must prevail over Emotion and that Rational thought must oversee all other thoughts and captions. In other words, one must treat humanity, whether in one's own person or that of any other in every case, as an end in itself and never as means to an end.

There is a principle that reason forms the fundamental principle of morality, thus granting very special status to mankind over other creations of God. The human being is bestowed with the power to reason, hence, the noble Quran calls him the *Ashraful Makhlooqat*, i.e., the most exalted one amongst all other creatures.

Kant also grants very special place to Justice.

Justice is an ongoing duty that requires granting an individual equal, unequivocal and his due as his right, without an obligation, favour or bias. It has two cardinal forms, i.e., Distributive and the Retributive justice.

Distributive justice is an important pillar of health care ethics. It has grown in importance as the resources are becoming less and less and the needs and demands of health care are progressively increasing by the day. Distributive justice involves a fair distribution of benefits and burdens in a society.

On the other hand, Retributive justice means matching punishment for a crime committed. It is not often enough that we see this form of justice applied in medical profession, unless a physician becomes part of a tyrant's arsenal to punish a so-called criminal, usually a prisoner of conscience, to inflict medical harm on such a person, labelling the atrocity as a form of punitive justice, as the tyrants normally do. Such an act on the part of a physician is to be strongly condemned and cannot be condoned for any plea including a threat to the physician's life or limb.

Justice is indeed a fundamental pillar of both religious as well as secular ethics. It is of immense importance in Islamic philosophy. Following the first attribute of *Tauheed*, *Adl* is the second most important attribute of Allah *Subahanahu Taala*. He is *Adil*, i.e., the Just One.

The noble Quran is full of texts, dictates and commandments on the principle of justice. No society can ever be called a just society unless the due is granted to the one who deserves. The divine dictates are quite clear about the fact that the society can only be just if it takes equal care of all its members, irrespective of the irregularities of birth or natural endowments. In fact, much is written in Quran in favour of the orphans, the destitute and the poor, who may otherwise be deprived of an equal status by the society. Islam is basically a religion that supports the poor and the weak. It strongly condemns any form of aggression or injustice, particularly warning against the usurpation of orphans' rights.

Justice is a tangible form of the conscientious thought applied in practice. It could be either individualistic justice or collective justice.

Individualistic justice has two fundamental dimensions, namely justice to one's own self or justice to another person, or a creature of God, such as an animal or plant.

No theory or a theorem can be called genuine unless it has its critics. Therefore, the theory of Deontology also has very devoted critics, who sometimes ruthlessly condemn it. One such avid criticism is that while deontology promotes goodness over evil and the means not being different from the end and so forth, it primarily focuses on the rights of an individual and grants unequivocal importance to an individual over the community. It also fails to consider the role of emotions and sentiments, affection and love, due to an overriding concern with duty and justice. After all, a man is not a mechanical device to possess only the supreme concept of the consciousness to the duty alone and base all his actions like a robot on the programming based upon reason and reason alone. There are emotions that every human being possesses, and one simply cannot be so clinical and so very abject in his life that he may disregard all other feelings and follow the command of observing reason and duty and justice only.

And that is the primary difference between all the Western philosophies that are mostly secular and the teaching of the Quran which is eternal and divine. Mankind is not a machine but is not infallible either. Therefore, the noble Quran has narrated many stories of people, both ordinary and exalted, that warn us against many errors that we often commit, in the name of justice, benevolence, empathy and duty. The Quran has clearly mandated that everything, every act must remain in balance, and any form of polarisation or imbalance would not remain good enough to be ethical. Allah has created everything according to a certain measure, i.e., *Qadr*, and nothing grows out of proportion, each following its own course. The moon must disappear, changing its shape from a thing of beauty when full to a pale and tiny shape and almost like the dying leaf of a palm tree, and the Sun must always rise from the East and so forth.

Duty is important in the eyes of Islam, but it duly takes into account the personal feelings, emotions and sentiments too. Besides, Islam clearly prefers community to an individualistic approach. Therefore, while it is mandatory to offer *namaz* five times each day, and doing so carries a reward, it is more rewarding if the same duty is observed in company, i.e., a mosque. Likewise, no doubt, service to one's family is compulsory and must not be neglected, but it is more rewarding if one could sacrifice the personal or family comforts to heed to the needs of more needy people of the community. And that is also the reason why so much importance is given to *Zakat*, *sadaqat* and charity. No one, indeed, in a classic Islamic society must remain wanting; and that is the fundamental approach of Islam.

Altruism and sacrifice are almost part and parcel of the Islamic teachings. One must give away what is not needed, but even better if one gives away even what he needs for someone more in need. The life of Prophet and his family is a glorious example for us to see a thousand examples of altruism in their daily lives. The concept of justice, i.e., *adl*, and benevolence, roughly translated as *Ehsaan*, is discussed elsewhere in an essay. Suffice it to say that *Ehsaan* is several steps higher than *adl*. As to the philosophy of sacrifice, the unmatchable story of Ibrahim and Ismael is sufficient to illustrate the significance of surrendering to the will of Allah and being ready to sacrifice the only son. The dream of Ibrahim was translated in a practical form by his grand son Hussain, A.S. in Karbala.

Deontology in many ways follows the principles of goodness and observing the code of ethics which appeals to common sense. It does not, however, condemn egoism or individualistic approach, which can sometimes be damaging to the whole philosophy of goodness per se. Islam clearly condemns self-appreciation, egoism, egocentrism, and the individualistic approach in all matters. It does not grant anyone special favours on account of any attribute other than piety or *Taqwa*. Islam believes in deontology, i.e., duty, but is more in favour of pluralistic and community-based service rather than a service to a person. The delicate balance between *Huqooq ullah* and *Huqooq* ul *Ibad* must be maintained in the light of Islamic teaching.

One must perform one's duty with honesty, integrity and sincerity and must grant more importance to the community than to one's own self; however, one must not ignore one's family either. Ali ibne Abi Talib said, "Treat your family with care just like a bird takes care of its feathers and wings, which protect it from the weather and *enables* it to fly high."

Deontology is a fine principle of ethics and grants proper emphasis to duty, justice, human dignity, autonomy, the rights and obligations and so forth. Islam of course teaches much more and remains the divine force that is growing in its impact across the globe as time passes. It is close to the human psyche and serves all its followers in matters of life and death.

Kant is famous for according great importance to human life. He abhors, indeed condemns, any act of violence, torture and suicide. He believes that suicide is an abomination, as it involves the misuse of our freedom of action to destroy ourselves and our freedom and autonomy, the very right to live and die.

All religions, particularly Islam, condemn the act of suicide. Therefore, to even consider an act of assisted suicide as something to ponder over is against the very thinking of both religious as well as secular philosophies.

The State of Oregon was the first to allow assisted suicide. Holland and some other societies in the West have actively encouraged this highly unethical practice. In Britain, the recent episode of Dr. Shipman, a practicing GP, was an eye opener. He killed hundreds of elderly patients in the name of helping them overcome their miserable lives. He himself committed suicide, which was perhaps an easy escape for him.

He believed, like many other physicians with a distorted vision of service to mankind, that in order to relieve one of misery, an act of mercy killing is justified. That, of course, is condemned and cannot be endorsed by any religious philosophy or by eminent philosophers like Kant.

They can't face the Kantian code of 'Universalising the moral principles,' as it says that if it is wrong to commit suicide, it always has to remain wrong to commit or assist such an act irrespective of the circumstances. Kant believes in the concept of 'absolute duty' irrespective of the prevailing conditions or changing circumstances.

Islam has given a clear mandate against all forms of violence, torture, inhumanity, disgrace to mankind, and any form of insult to one's individual freedom. Suicide, of course, is strongly prohibited in Islam.

Now let us discuss another important ethical theory. It is called 'Utilitarianism.'

According to this highly respected and universally acknowledged theory, 'the morality of an action is determined solely by its consequences.' Such actions that may bring happiness may therefore be called morally right, and those that may bring unhappiness, pain, discomfort, harm or agony, obviously are unethical and immoral.

The theory of Utilitarianism grants more importance to the overall interest of the community over an individual. It therefore has a distinct advantage over the theory of Deontology, which primarily focuses on individual.

Jeremy Bentham and Stuart Mill, two great philosophers, promoted this theory as a way of promoting and achieving social reform. And in many ways they did succeed, though not without their own share of condemnation, criticism and outright disapproval.

The theory of Utilitarianism is often unclassified into a duty to promote good (Beneficence) and to do no harm (nonmalificence), which in Indian teachings is called 'Ahimsa.' These two principles have been incorporated by Beauchamp and Childress in modern medical ethics as Beneficence and Nonmalificence, two of the four pillars, as described by them.

Bentham has gone into considerable detail to list seven factors that should be considered and employed as the measuring tools in evaluating total amount of 'pleasure' resulting from an 'act.' These factors are Intensity, Duration, Certainty, Propinquity or nearness in time of pleasure or pain, Fecundity or how productive the pleasure will be in promoting goodness, and finally Purity and Extent.

The importance of Utilitarianism rests with its practical application in today's world of scarce resources, often used by the health care pundits in evaluating cost/benefit ratios in the rich countries. The question of rationing is nothing new to the rest of the world, who are simply in a state of constant rationing, grateful for any benefits they might get in an otherwise moribund society.

As we discussed earlier, while Kant's concept of Deontology attaches singular importance to human dignity, and respect of the individual's rights, the Bentham theory of Utilitarianism utterly fails to do so and that too with a purpose. It promotes the concept of sacrificing an individual's needs, rewards benefits even happiness for the greater benefit of a community or the society. Therefore, in the fateful 1930s, the Nazis employed this theory in carrying out mass scale human experimentation in their concentration camps, wrongly justifying their heinous acts as scarifying a few for greater benefits to mankind. For obvious reasons it brought a lot of ill will and bad name to the theory of Utilitarianism, which to say the least, does not allow any form of wrongdoing to mankind, even if for the larger good of the community.

The Nazis, under the command of Hitler, Himmler, Goering, Goebbels and their likes, executed the Jews and the gypsies as well as mentally disabled people of all ages and both sexes in their damnable camps at Auschwitz, Treblinka, Dachau and elsewhere. The name of Dr. Joseph Mengle, for carrying out human genocide in the name of research, employing the theory of Utilitarianism to his unethical motives, is condemned in the annals of history of medical ethics.

The Nuremberg trials after the Second World War opened the eyes of the public and challenged their conscience on the atrocities carried out in the name of scientific research. The name of Nuremberg has forever been etched out on the annals of bio-ethics through the famous Nuremberg Code (1947).

Regrettably, though, human experimentation has not ceased to be carried out. Most of it is now ethical as the Nuremberg Code dictates morality, but many unhappy episodes of unethical scientists come to the world's attention every day. Nazis were not the only ones doing atrocious human experimentations in 1930s, as we now know that American scientists had begun their infamous Tuskegee experimentation on innocent and healthy black people of Birmingham, Alabama in late 1930s, too. Their atrocities, however, remained masked until as late as 1974, when a whistle-blower blew the whistle. And even then, the information remained confined to the few until 1997, when American leadership had to apologise to the victims' families on national television. Thus, the whole world came to know of those horrible tragedies of modern-day America.

Unfortunately, no theory of ethics can guide or control an evil mind. A scientist who is determined to abuse his power granted through knowledge and expertise through his technical achievements can at any time distort or concoct a theory to suit his motives. Such people have no conscience, which could otherwise control their actions. The world is full of such evil minds, which unfortunately control the destiny of many ordinary human beings. That is why we must expose the evil acts performed by these people and continue to propagate the knowledge of medical ethics. The only way to somehow eradicate such atrocities from happening again and again is by putting up strong and forceful resistance against the evil forces.

There are many differences between Deontology and Utilitarianism. But the whole thing can be summarised by saying that in Deontology, the 'means as well as the end' must both be good, whereas in Utilitarianism 'an end may justify means'. Besides, Deontology grants precedence to an individual and his rights whereas Utilitarianism strongly prefers the benefits to the community over an individual.

Islamic teachings are quite obviously far better since without compromising the rights of an individual, they grant special importance to the role of community and the society at large.

There are many forms of *Ibadaat* that a Muslim must perform in his life time. Some are mandatory such as *Som- o- Salat- o Zakat*; others are optional such as *Sadqah* and *Khairat*. But these are individual obligations. Many others are communal in nature and are more important from the point of view of closeness to God. One such *Ibadat* is the service to the community. Consequentialism may be justified in some situations as *'Maslaha'* and may justify the means that may not be good enough but may end up with a positive, fruitful and beneficial nature to the community.

All religions advocate charity and service to the community and consider it to hold a higher moral ground compared with individualistic worship. While Christianity believes and practices celibacy and hermitage, Islam disallows both. The former may lead to unethical and therefore irreligious sexual practices, as was seen in no less a place that the Vatican of the Borgia popes; the later may lead to solitude and deprivation.

Morality is adored by all. Irrespective of beliefs, most religious as well as secular societies agree that to be moral means to be good. Nearly all philosophers in the West except Miaccahaevelli have condemned abuse of power and authority by the rulers or the state. He was the only evil philosopher who promoted oppression of the poor and helpless through his adverse code of ethics, namely 'rule them with fear,' which has not been popular except by evil rulers.

Morality is a likeable attribute. Even those who are evil would someday approve of a moral person as the ideal person in an otherwise evil society. These tyrants may kill, maim and torture such pious people, as we have seen in human history, but they can never call themselves moral by condemning such a person despite the hatred for him. Moral values are universal albeit deeply influenced by cultural norms, beliefs, faith and many other variables. No one, not even a tyrant like Saddam, could call bravery, or generosity, or justice, or truths anything but goodly attributes i.e., virtues. And one may suspect as was seen in the trial of Chemical Ali that Saddam did indeed believe that he was a fair and just ruler and that his action against the helpless Kurds and the Shia population was in the larger interest of the Iraqi. And that is an absolute lie. Such are the situations where tyrants and evil people have applied the Utilitarian theory in their evil actions, justifying their evil actions in the light of philosophy. How very untrue, unjust and evil that could be.

Virtue ethics is another important ethical theory. It concentrates on the character of a person, not the actions. Aristotle claims that a virtuous person thinks and practices virtuously. By nature he is made that way and is obliged to act accordingly, as it is a part and parcel of his nature.

In fact, logically speaking, Aristotle's virtue theory appeals to the common sense. It had lost its glory over the millennia, but thanks to MacIntyre the virtue theories are being revived, as are Neo-Virtue theories. Aristotle was a scientist and a deep-thinking philosopher. His approach was very practical and precise. He was wrong on many themes and did indeed support slavery, for which he and other Greek greats cannot be forgiven, but undoubtedly his Virtue theory is absolutely fascinating.

Aristotle believed in the inherent goodness of some people and he called them virtuous people. He believed that such people cannot conduct an evil act as their system, i.e., soul or the spirit or the conscience would not allow it. So, such people are incapable of doing an evil act. Was Aristotle saying that these people are bestowed with divine powers to act only for the good or was he saying that such people have a strong moral code embedded in their DNA that determined their actions per se?

In the Shia faith there is the concept of '*Massomiat*.' It basically means that certain chosen people of God have a built-in mechanism that makes them perform only the goodly acts. In other words, they cannot perform an evil act. They are infallible. Perhaps even more appropriate would be to say that they are by nature and by their deeds 'good' in all respects and an evil act, i.e., sin, cannot touch them.

The word 'Asmat' is the root word for *Masoom*. *Asmat* basically translates as piety and its adjective is *massomiat*. The person who possesses *Asmat* is called *Masoom*. In the Shia faith, all the Prophets were *Masoom* (infallible). Prophet Muhammad SA and the twelve Imams are all *Masoom*. They were virtuous people who were bestowed by Allah the divine attribute of Asmat and were therefore called *Masoom*. They were, in fact, the lighthouse for the *ummah* to judge their actions against the parameter of the acts of the *Masoomeen*.

If Aristotle believed in virtuosity and called some people to be virtuous by nature because they possessed the inherent attribute of goodness in them, then one may say that the concept of inherent virtuosity or '*massomiat*' is not new, and must have existed in other cultures way before times. One Masoom, i.e., infallible imam, Ali Ibne Talib, gave a sermon on Faith or *Yaqeen*, partly quoted in an essay here. But the finest piece of his teaching may be witnessed in an epistle that Ali wrote to his subordinates on ethical principles and good governance.

That epistle is a masterpiece on work ethics and is included here in the form of an essay. It is mind-boggling to note that despite the lapse of nearly a millennium and half, his words of wisdom appear to echo in the halls and corridors of the palaces of the mighty rulers and Governors, as if time has stood in an extended moment of suspension, and the world stood motionless listening to Ali as he spoke.

Josh in his unique style wrote

Khamosh! bey adab key Ali bol rahey hein.

2.3 Faith: A Virtue

Aristotle *has defined Knowledge as Virtue. Is faith also a virtue?* Let us discuss it.

Kant's philosophy of Deontology is based upon the principle of human dignity and autonomy. It is also based upon the fundamental rule of doing a good act with the intentions of bringing about goodness as a consequence of such an act. He believes that an act should be a means as a well as an end, to achieve the desirable good results. In other words, the intentions as well as the deeds of an act must both be good and be justified to be called deontological.

In modern medical ethics, as in many other fields involving human beings, nothing is more precious than human dignity. The autonomy and the individual's rights must remain supreme in all matters of life and death. Of course, since the multitude of individuals form as a community, when it comes to the concept of collective human dignity, the rights of all the members of the community must receive equal and unbiased status.

In the olden days of ultimate human indignity and disgrace, when slavery was a common practice, the ultimate humiliation that a man could sustain upon fellow human being was the total control on his subject's life and death. It was an era of total disregard for human values and the rights of an individual or community. The end of that era is never to be forgotten by those races who suffered the indignity for generations. With a visit to Liverpool or Bristol, two of the infamous slave cities, one can still see the concrete engravings and figurines of the African slaves chained and shackled and forever plastered onto the walls of old colonial buildings as a constant reminder of those horrible days.

In the medical profession, human dignity is taken into account not only in matters of medical care but also as a marker for the most fundamental rule of medical care called faith.

Faith is indeed the cornerstone of the medical profession. It has stood the test of time and has remained supreme in all matters of medical care, over the millennia.

The principle of faith and trust is the binding force that has kept the doctor-patient relationship on an even keel. It is a bond of love and understanding. It is a sacred covenant that often remains unsigned in most medical professionals but has very strong undertones in day-to-day dealings with the patient. A violation of this covenant is almost like demolishing the very foundation of a relationship, resulting in many untold tales of unhappiness. Maintenance of this sacred covenant and keeping it solid is an ethical act per se; its violation is totally unethical to say the least.

The patient comes to a physician in total confidence and with total faith unto him; hence, it is obligatory upon a physician to not only uphold that covenant of trust but to bolster it with his acts of benevolence, kindness and genuine feelings of care and attention to his needs. How true is the time-honoured statement used in Indo-Pakistan that a kind word to the patient is a better healer than an apothecary's lotion or potion.

Much has already been written on the subject of faith and trust between a physician and the patient, but it is an ocean of unfathomable depths, and every swimmer dives deep to his capability to seek the pearls of wisdom, and to benefit him and share it with others. Such is the joy of research and writing.

From the days of the cave man, illness has been considered to be evil. Often enough it was considered to be outcome of an individual's evil deeds, but also the result of the evil spirits casting unholy shadow on the sufferer. The demons and the ghosts have been the part and parcel of many a folk tale. They continue to remain a strong force even in today's world, when the world has changed into a global village closely knitted through the global explosion of info and technology. Lest we forget the power of the misguided and misguiding *pirs* and *fakirs*, even today one can see the *Daulay Shahkey choohey* in many villages and some cities of Punjab and Sindh. They are often seen chained and shackled to the master who treats his subordinate as little more than a chimpanzee.

Sigmund Freud, highly controversial but extremely potent master of human psychology, changed the theosophy of the so-called hysteria only few decades ago. Until he could convincingly put forward his theories of dreams and sex, an important force behind many psychological and psychiatric illnesses, many unfortunate young girls were left to rot in asylums. This great philosopher of human mind owes much more credit that he has been given so far, as he must have saved thousands from the ECT and condemnation to torture. All such patients were considered to have been spell-bound by the evil spirits.

In our villages, it is not uncommon to witness a young girl suffering with hysteria, to be tied down to the *charpoy*, a pillar or a post with a rope, and subjected to the *jadoo* mantar of an illiterate but self-proclaimed authoritarian mullah, deeply absorbed in an act of exorcism, while making a small fortune at the cost of the unfortunate woman.

The Western media has also not stopped showing the films showing such awful and sickening events, and the writers have not stopped writing about evil stories of the demons and bad spirits. They are sick to say the least, if not outright mad, requiring psychiatric treatment if not institutionalisation, to help them overcome their intoxication with evil.

All that is the result of ignorance, the tangible form of which is *shak* or suspicion. While *shak* or suspicion. i.e., the lack of trust, is a sickness of the human spirit and souls, knowledge and faith are the cures for such an illness.

Faith unto your physician and the treatment given by him is a lesson taught to the patients. And that the patient is like your own family member is the lesson that physicians have been taught since the days of Galen, aka Hakim Jalinoos.

In an Islamic culture, faith unto Allah is the fundamental teaching that we all have been taught since our earliest days. He alone knows the illnesses that you suffer and Quran, 26:80 He alone grants you the remedy [1]. Allah alone is the *Aalim al ghaib wal shahahdat*, and He alone knows of the future happenings. No one but He alone can save one from a disaster, and to Him alone we all have to eventually return. He has created us from nothing and He shall return us to nothing.

Learned Islamic scholars have defined three categories of faith, i.e., *Yaqeen*. The first category is called Ilm al *Yaqeen*, i.e., you see smoke and believe that there must be a fire. The second category is called *Ain Al Yaqeen*, i.e., you see a fire yourself and need no further evidence. The third category is called *Hadd al Yaqeen*. It means that you jump into a fire and have its impact on yourself.

The first category is a common observation that we all have. We see the signs of Allah's creation and have firm belief in His existence. The second degree means that one experiences and closely observes Allah's existence and His total control on everything to be closer to Allah. That is what Ali Ibne Abi Talib said when asked if he saw Allah. He replied, "I do not worship an Allah whom I do not see." The third and final degree is the ultimate state of faith, when Ibrahim jumped into fire at the command of Allah, and Hussain sacrificed his life and those of the dear ones at Karbala as a duty to his grandfather's faith and preservation.

Medical care in a God-fearing society is based upon the trust in Allah and trust unto yourself for treating the patient whose charge has been given to you. Faith has to be the basic ingredient in all matters of health care. Faith unto your knowledge and skills to treat the patient, faith unto the powers of drugs or intervention employed by you, faith unto the ultimate healer and faith unto the mortality of mankind. And also faith unto the art of your profession that has granted you such immense responsibility that you must honour with beneficence and with nonmalificence.

Faith goes hand-in-hand with prayers. How often have we not seen that in hopeless situations, nothing but faith unto God and your prayers come to the rescue? Islam clearly defines the limitations of the mankind, in all matters of life and sickness. Allah alone grants us the joys of life as well as sometimes hardships, for which He alone gives us the strength and courage. Prayers are the medium through which we appeal to His authority for mercy and forgiveness. He listens to our prayers, as He is everywhere and omnipresent. In the Quran, Allah has clearly mandated that He is closer than the *Hablal Vareed* or the Jugular vein.

The role of a physician is to do his best to ease the pain and discomfort and help the patient recover from an illness. And that must be done keeping in view that Allah is closer than your Jugular vein and observes all your acts, which are for the matter of record constantly being updated by two scribes called *Karam- al Katebain*.

In the ancient Jewish tradition, the importance of faith and prayers is witnessed in the media, when the orthodox Jews are shown religiously and faithfully reading the Torah while facing the Wailing Wall in Jerusalem. Moses Maimonides was the finest physician in Moorish Spain. He was a Jew and prospered in Andalusia between 1135–1204, where he observed his faith diligently and served the mankind with utmost faith in the Almighty. He wrote down the physicians' prayer, which has stood the test of time, and continues to remain a part of the Jewish culture of faith and trust in all matters of healing and sickness that only God (Jehovah) possesses. Only recently, though, the credit for Maimonides physician's prayer is being transferred to Marcus Hez, an eighteenth-century Jewish physician.

In the Christian faith, the finest illustration of a physician's prayer can be seen in the form of the *Gospel of Luke*, who was himself a physician by profession. The followers of Jehovah's Witness, the Amish living in Pennsylvania, and the followers of the Latter Day Saints have all firm faith in an omnipotent God, and their physicians duly put their faith in His healing powers when treating their patients. They pay regular visits to their temples and other places of worships. In fact, the Amish believe so strongly in God that they often decline any form of intervention in their medical problems, which is exactly in accordance with our faith in matters of health and sickness. We are ordained to put in all our efforts in the care and management of our sick, though the outcome is not in our hands.

Some years ago I read a story in an illustrious magazine about the faith of a brain surgeon. When all his efforts to control haemorrhage in a surgery for a tumour failed, he sat down in a corner asking his assistant to keep the pressure on at the bleeders, and quietly prayed to the Lord with utmost humility and total faith. He said that no sooner that he had asked for forgiveness and pleaded for mercy, the Lord answered his prayers and the haemorrhage stopped. His prayers were answered. Miracles happen every day. Some we can attribute to human efforts, others are simply inexplicable.

It is a matter of day-to-day happenings in any doctor's life that we often feel helpless and hopeless about in some situations when our faith unto God rescues us and saves us from untoward happenings. A faithless person may not agree, but in his solemn moment of adversity, I am sure even he asks for help from the ultimate source of strength, whatever he may call him.

In India, the Hindu *mat* is deeply immersed in the philosophy of meditation and prayers in the temples as well as in solitude. Their philosophy is a very basic one, that the Omnipotent God has supreme powers and all human beings have to submit to him alone. The ancient Vedas and books of *Ayvervedic* are inundated with the humans and praises for the Creator, whom the Sikhs, just like Moslems, address as *Rab*.

Pakistani culture has two clear aspects to witness. Everyone has a firm faith in God, but some believe in the saints and *pirs*, and submit to God, pleading for help with the reference of the saints, while others prefer to communicate with God directly. It is all a matter of faith and belief. Even for a job we need references, so for praying to God, to plead for our case, we certainly need the best possible references. Or for some it could be the *pirs* and fakirs, for others *Mohammad o Aley Mohammad*. Obviously the better the reference, the higher the chances of approval of your plea!

At a visit to Hazrat Abdullah Shah Ghazi's mazar at Clifton in Karachi or the Data Darabar in Lahore or many other similar places of Dervishes and Saints' mazars, you can see for yourself the miracles that happen day in and day out. Or else, why are these places inundated with throngs of people, who obviously receive something from these places that they keep returning to them?

I have personal experiences of my own when I have pleaded to Allah at the threshold of Maula Abbas AS in dire situations with full approval of my pleas, not once, not twice, but ample times. He is called *Babul Hawaij*, i.e., the doorway to the solution of your needs, approval of prayers and grant of largesse by Allah ST.

The famous church in Lourdes in Italy is yet another example where Christians from all over the world take their sick and dying for the blessings of the Lady Fatima, and where hundreds of people gather night and day for the mercy of God. Many of them return as healthy as they were before even as deadly a disease as cancer afflicted them.

All cultures and civilisations, from times immemorial, have placed their faith in the Lord God for the solution of all their all, particularly the matters of sickness and disease.

From the days of the Borgia popes and later on, the periods of medieval Europe are full of strange of tales of abuse of power by the Church and the clergy in the name of religion and faith. For hundreds of years, young women suffering with many psychiatric illnesses were condemned to death on the gallows, when they were stoned to death in the name of ridding their souls of evil spirits. But the abuse of religious faith is not confined to any particular religion—it is almost universal. Haven't we heard of the village mullah misguiding his simple followers of the mystical powers that he possesses and how he can look into the future? No one but Allah has the knowledge of things to happen and times to come. Such is our belief on the attribute of Allah, i.e., *Aalim al ghaib wal Shahadat*.

Despite the wealth of money and resources put into the Church, unfortunately the British society has become almost godless. Only a few people go to church and fewer still believe in the Omnipotent God. The disregard for religion has made many of them rudderless. Authority, once eroded, seeps through the entire fabric of the society, and that is why there are so many stories to be told, whereas the British themselves see the problems but can't do much about them.

Faith is good for calmness of the soul, hence of the body. A person who has faith is often seen as calm and sedate, neither bothered about praise nor any rebuke. Because he has faith in a supreme power that determines his destiny, he is at peace with himself; doing his best to achieve certain goals in life, he has full faith in the principle of *Qaza o Qad*r. What will happen shall happen is his motto, but that does not deter him maintaining his cause on target and with a purpose, and usually his goals are simple and easily achievable. And his manner is simple and his exterior devoid of any show-offs or displays, his inner self content with his life and its achievements. He is modest and simple in living, and yet his intellect and the wisdom may surprise the onlooker.

In Russia, the Church was very strong. After the Bolshevik revolution of the early twentieth century, the churches were closed down so that the people would become totally dependent upon the communist authorities, as the hope and the guidance they obtained from the church was ruthlessly snatched away. Men like Rasputin, therefore, exploited the simple masses of these people who had total faith in a healer above, but when deprived, they turned in hope to this evil person.

The use of talisman and sooth sayings is part of Asian culture. It is only a way of expressing one's faith in a supreme being who helps them in sickness and in grief and saves them many a misery.

Faith is the pillar of ethics; just the lack of it is akin to having a building without a supporting pillar.

The philosophy of Faith, i.e., *Yaqeen*, is best explained in *Nahaj Al Balagah* by Ali Ibne Abi Talib.

Imam Ali said' Faith is like a building, supported by four pillars, namely Endurance, Conviction, Justice and Jihad. Endurance is composed of four attributes: Eagerness, Fear, Piety and Anticipation (of death). So whoever is eager for the heaven will ignore temptations, whosoever fears the hell fire will abstain from sins, and who so ever anticipates death shall hasten toward good deeds'.

He went on to explain it further: 'Conviction also has four aspects: to guard against infatuation of sins, to search truth through knowledge, to gain lessons from instructive things and to follow the precedent of the past people, because whoever wants to guard himself against vices and sins will have to search for the true causes of infatuation. And the true ways for combating them and to find those true ways, one has to search for them with the help of knowledge. Whoever gets fully acquainted with various branches of knowledge will heed lessons from life. And whoever learns lessons from life is actually engaged in the study of the causes of rise and fall of the previous civilisations.'

Ali went on to add in this sermon 'Justice also has four aspects, namely, depth of understanding, profoundness of knowledge, fairness of judgement and clarity of mind. Anyone trying to achieve these traits will have to develop ample patience, tolerance and forbearance; and whosoever does just this has indeed done justice to the cause of faith and has led a life of good reputation and fame.'

Finally, he said, 'Jihad is divided into four branches, too, namely, to persuade people to be obedient to Allah, to prohibit them from committing a sin, or vice, to struggle sincerely and firmly on all occasions and to detect the vicious.'

The reader is advised to read the entire sermon in the English translation of *Nahjal Balagha* by Syed Ali Raza.

Akbar Ahmad is a renowned social anthropologist and a scholar. He has written on many matters related to culture and its impact on human life in many of his essays. One such article was written by him a few years ago on the concept of faith.

After the tragic events of a family from rural Punjab, in Pakistan, who were drowned in the Arabian ocean, several people wrote on the event, including Dr. Ahmad. These simple folks had firm faith in their religion and its dictates. They came from simple, Shia families. It was their utmost desire to travel to Karbala and Najaf e Ashraf for pilgrimage. They were poor and had no means to do so, but they had full faith in their belief that one day they would receive a divine call to reach Karbala. Therefore on that eventful day, in their solemn belief, they all walked onto the waves of the Arabian seas. Some of these believers drowned and others were saved and sent there by a philanthropist on the pilgrimage, but certainly they all reached their goal, spiritually if not physically. Metaphysics teaches us about many a supernatural phenomenon, which is difficult to explain but does happen and continues to pose a challenge to human intellect.

After all, it was the faith only that had made Abraham jump into the *aatishe Namrood*, and it was the trust that Ishmael had in his illustrious father's dreams that he surrendered to the seemingly ungainly and rather frightful command of Abraham to lie down on the slab for a sacrifice to please God almighty. How true is the commandment of Hussain Ibne Ali that it is far better to die in faith than live in doubt?

Faith displays itself in all cultures and in many colours. It is the firm belief in what you believe to be right that grants you the courage to perform what you would ordinarily dread. It does not have to be the Muslim religious faithful only who can walk barefoot on the burning ambers; it could well be a monk of Buddhist faith or a savant of Hindu mat. Both can surprise an onlooker with similar acts which they perform purely on the firm belief in the act itself and the strong motivating force behind it.

In the Native American culture, the concept of healing, much like the ancient Chinese, the Incas and the Aztecs, revolves around the notion of four powers, strengths and weaknesses, of the human beings. They are physical, emotional, social and spiritual.

They believed that most illnesses are the outcome of good or evil attitudes of a person, individually or collectively towards the family and friends, peers, seniors, and those who matter in their lives. Thus, they believe that the delicate balance between the emotional and spiritual well-being on the one side and physical and the social well-being on the other maintain the status of health. Any imbalance in these components, and one may end up with either physical or emotional ill health. It is the job of the wise man, the *ved*, the Hakim or the *Tabeeb*—in Native American culture is called the Shaman—to advise his people to maintain the balance between the components mentioned, and if afflicted by sickness or disease, the ways of escaping a disaster.

Healing can only take place if there is a mutual, genuine and sincere understanding between the Shaman and his patient. Obviously the Shaman must conform to the code of conduct that is expected of him as a wise man in which the layperson has laid his trust. That covenant of trust and faith is beyond all religious beliefs and is universal in its application. Since healing is also a spiritual process, the character of the Shaman is all-important in achieving a good outcome, according to the belief of the Native Americans. He must not have false pride, grandeur, haughtiness or reck-lessness in his dealings with the patients, as indeed in his ordinary life. In many ways, he should be a role model for the community; that is one reason why he is granted the status of a wise man, sitting on a higher pedestal.

India is a subcontinent which can safely boast upon its ancient tradition of Ayurvedic medicine, its Upshinaiads and respect for the human body and soul, through the discipline of yoga and meditation. It can also safely boast about the diversity of its culture, myriad religions and faiths, and multitudes of population ranging from the mountainous inhabitants of the Himalayan territories to the plains of Rajasthan. There are more Muslims now in India than Pakistan, and their president and the top scientists, as indeed many politicians, are all Muslims. Besides there are, of course, Hindus, Christians, Buddhists, jains and many more with a vast array of faiths and disciplines.

The ancient text in Hindu medicine is called *CARAKASAMHITA*, which was documented in the first century AD. The word *Ayvervedic* embodies the literature on medicine derived from *Carakasamhita*. [2]

Ayvervedic is deeply dependent upon religious faith, or *Dharma*. It involves not only the religious beliefs, but also spiritual awakening. Ayuvereda is a combination of a trio of health, religion and morality. In the ancient scripts of Hindu literature, it is quite aptly described that right and moral conduct can provide a degree of safety and security from disease and sickness. The teachings of Karma strongly advise that the moral behaviour involves not only observance of physical goodness of health, good food, and proper rest and sleep, but it also involves avoidance of such evil elements as greed, anger, malice and jealousy.

Ayvervedic medicine does not separate body from mind. It believes that an individual's health or the lack of it is often dependant on his mental state, thought process and spiritual disposition.

In Hindu medicine, i.e., Ayervedia, the physician, called Vida, enjoys a high position in the society, perhaps second only to the priest. It is clearly emphasised in the ancient literature that a Vidaya can only be helpful in serving the sick if he has no greed, avarice and self-indulgence. His primary aim must never be material gains or elevated status in the society and never be based on false values not to save the mankind in a humble way. He is only a physician doing his best to ease the discomfort of the patient, and it is the Omnipotent God who will determine the final outcome of his endeavour.

The Vadiya must be honest, truthful, knowledgeable and skillful and should practice his art with humility, modesty, justly and in total confidence. He must not betray his patients' faith unto him and must know his limits and bounds. The highest level of health in Hindu *mat* is the level of achieving spiritual enlightenment called *Moksha*.

There are four aims described in the Hindu *mat* likewise to achieve health and happiness, and they are

(a) Material comfort—*Artha*
(b) Right conduct—*Dharma*
(c) Physical pleasure—*Karma*
(d) Spiritual awakening—*Moksha* [3]

It is expected of an Ayvervedic physician or Vidaya to practice all those elements, more so the moral (Dharma) and Spiritual (Moksha) to best function as physician—a role model—in whom the patient can put his faith in and trust with the most valuable possession of his life, i.e., health.

Therefore, it appears that in most societies across the world, trust, confidence and faith play an integral part in the art of healing. It is not simply the physical ailment for which one is seeking remedy but also the spiritual cleansing and emotional uplift which are all integral parts of the well-being of human beings. In today's world of holistic approaches to medical care, it is not adequate enough to treat an organ or a system alone but the entire human body including the care of the environment and other factors relevant to the illness.

In the process of holistic medical care, the physician has to be more knowledgeable than his colleagues of the past. He not only has to have in-depth knowledge and supporting skills of his trade but also good knowledge of the social sciences, psychology and the living conditions of the patient. Ideally, he should know sufficiently about the factors that may be causing or contributing towards the patient's illness, and not just the clinical symptomatolgy alone. Medicine today is holistic.

Such a physician may thus contribute towards the regaining of health by the patient, both physically and spiritually. In this process of achieving this objective, faith plays an integral role. Without the patient having faith in his physician, and without the physician having full faith in his knowledge and skills, total success may not be achievable. In triangular equation of educational psychology, besides knowledge and the skills, the third element often described by the pundits as attitude comprises such intangible elements as etiquette, bedside manners, mutual respect and care, love, trust and faith. All these elements combined together may be labelled as Ethics. A perfect harmony between knowledge, skills and attitude is what may bring about a visible, noticeable, palpable, tangible and constructive change in a physician. And that is also the definition of Knowledge, an objective of medical educationists the world over.

If there is any form of imbalance in the triangle of education mentioned above, you may not end up with a desirable finished product. Too much knowledge without supporting skills may give you an academic to sit in the lab and twist his grey cells to squeeze out the juice but may not give you an ideal physician. Likewise, lack of knowledge without a solid foundation of skill will almost invariably give you an artisan a *mistri*, not a maestro. This is like a technician, a cutter but not a surgeon who would know when not to use his knife or where to draw a line. In a university hospital in the UK, one surgeon called himself a cutter, for his personal reasons, I suspect due to his lack of interest in education or research, only surgical procedures!

Similarly, if you have these components and the person is devoid of manners and ethics, he is often described as 'a person with an attitude,' i.e., not quite proper in matters of patient dealing or dealing with his colleagues, peers or seniors. Such a person is often a lopsided individual boastful of his expertise, without understanding human nature or the delicate balance that it possesses with many other components. Such a person would almost fail to inculcate faith and thus fail to deliver the desired results in patient care.

Faith grants confidence to both the patient and the physician. Its presence is a sure sign of balance and success, the lack of it means failure.

Knowledge is a virtue, which enables one to differentiate between the good and the evil. Faith is an equally meritorious virtue, which grants you the power to make definitive decisions based upon the principles of trust and confidence. That is why of all *momineen*, the holy Quran has granted singular importance to the status of *Rasikhun fil Ilam* (*Immersed* in knowledge), who have ultimate faith, i.e., *Yaqeen-e- kamil*.

2.4   Work Ethics. Inter Professional Ethics: Imam Ali's Directives

The current global financial meltdown in major countries of the Western world is caused by a financial crisis that was brought upon us in 2006–2007. One scholar pinned it down to a simple phrase when he said, "The bankers broke their code of ethics and destroyed the global economy." He is absolutely right. If only the bankers had kept our faith and observed the very basic code of ethics, namely trust, we should have been all right.

Not only that, but they also became greedy by gambling with our deposits and losing billions through speculative banking. Hundreds of thousands of ordinary folks, pensioners and widows lost their lifetime savings and are now struggling to live just above the poverty line in the UK, Europe and America.

The oft-repeated statement of the celebrity financial analysists and TV anchors has become a regular feature for a common man. And that is the lack of observing Work Ethics.

The medical profession is inundated with the stories of petty jealousies and politics that go on with in the profession and between the colleagues. It is a global phenomenon but most prominently encountered in the premiere institutions and the so-called Ivy League centres. Everyone is trying to elbow the other out. As Churchill said in Parliament, "My foes are in front and my enemies behind." How true. Every Caesar has a Brutus lurking around in the shadows. One sees it all the time in the medical profession, which is perhaps also true for other professions and trades.

Glorifying one's attribute, through display of degrees, often fake or cooked up, happily printed out in large bold prints is a common scene in the Indian subcontinent and many other developing countries: The bigger the publicity board, the more qualified the doctor, appears to be the motto. Many fake *hakims* and *veds* use other outrageous ways of highlighting their expertise to the ordinary, gullible village folks.

Backbiting, whispering campaigns, maligning colleagues and sycophancy are some of the attributes that the medical profession suffers with. Highlighting one's own capabilities against a fellow colleague or belittling him is a customary practice in many countries. It is a universal and international ailment that the noble profession practices. So we need to learn a lesson or two on professional and inter-professional work ethics.

Numerous codes of work ethics have been published over millennia by the law-giving agencies and leaders. Every philosopher has written volumes on codes of ethics for people to follow, but the greatest of all philosophers best described by an illustrated journalist as the 'philosopher- soldier' of Islam, Ali Ibne Abi Talib, wrote a mandate on work ethics which has stood the test of time.

Ali was the Ameerul momineen, *Imam e awal* and fourth caliph of Islam. In that capacity he commanded his Governors of different provinces under his domain. This Epistle was addressed to Malik e Ashtar, the Governor of Egypt, which is amazing, practical and based upon the principles of humanity, dignity and honour, the universal principles of mankind.

I have included it here to learn many lessons in work ethics as a common man, as a professional, as a person of authority, an administrator or medical director, even a chief of Health services, a minister or a Governor. In fact, the present Governor of Sindh is a doctor, and many of my students are parliamentarians, politicians and ministers. The currently-under-siege President of Syria, Bashar al Asad, is an ophthalmologist, as was the founder of modern Malaysia, Mahatir Mohammad. A couple of decades back, Sukarno, the revolutionary leader of Indonesia, was also a physician turned politician.

If only they had read this document of Imam Ali, and applied the instructions given in it, their jurisdiction would become a role model for others to follow!

Here is an extremely authentic translation by Allama Rasheed Turabi, an extremely renowned orator, speaker, writer, Alime din and Zakire Ahle Beit, of the famous Epistle written by Imam Ali to Malik e Ashtar. It is available on the internet.

According to the historian Masudi (Murooj-uz-Zahab Masudi Vol. II, p. 33, Egypt), Hazarat Ali is credited with not less than 480 treaties, lectures and epistles on a variety of subjects dealing with philosophy, religion, law and politics, as collected by Zaid Ibn Wahab in the Imam's own life time. So highly valued are these contributions both for their contents and their intrinsic literary worth that some of his masterpieces have formed throughout the course of Islamic history subjects of study in centres of Muslim learning. Indeed, his reputation seems to have traveled into Europe at the time of the Renaissance, for we find that Edward Powcock (1604–1691) a professor at the University of Oxford, published the first English translation of his *Sayings* and in 1639 delivered a series of lectures on his *Rhetoric*.

This letter according to Fehrist-i-Tusi (p.33) was first copied in the time of Hazarat Ali himself by Asbagh bin Nabata and later on reproduced or referred to in their writings by various Arab and Egyptian scholars; chief of them being Nasr ibn Mazahim (148 A.H.), Jahiz Basari (255 A.H.), Syed Razi (404 A.H.), Ibn-i-Abil Hidaid and Allama Mustafa Bek Najib, the great scholar of Egypt. The last named regards this letter 'as a basic guide in Islamic administration.'

Ali wrote: 'Be' it known to you, O, Malik, that I am sending you as Governor to a country which in the past has experienced both just and unjust rule. Men will scrutinise your actions with a searching eye, even as you used to scrutinize the actions of those before you, and speak of you even as you did speak of them. The fact is that the public speaks well of only those who do good. It is they who furnish the proof of your actions. Hence the richest treasure that you may covet would be the treasure of good deeds. Keep your desires under control and deny yourself that which you have been prohibited from, for, by such abstinence alone, you will be able to distinguish between what is good to them and what is not.

Develop in your heart the feeling of love for your people and let it be the source of kindliness and blessing to them. Do not behave with them like a barbarian, and do not appropriate to yourself that which belongs to them. Remember that the citizens of the state are of two categories. They are either your brethren in religion or your brethren in kind. They are subject to infirmities and liable to commit mistakes. Some indeed do commit mistakes. But forgive them even as you would like God to forgive you. Bear in mind that you are placed over them, even as I am placed over you. And then there is God even above him who has given you the position of a Governor in order that you may look after those under you and to be sufficient unto them. And you will be judged by what you do for them.

Do not set yourself against God, for neither do you possess the strength to shield yourself against His displeasure, nor can you place yourself outside the pale of His mercy and forgiveness. Do not feel sorry over any act of forgiveness, nor rejoice over any punishment that you may mete out to anyone. Do not rouse yourself to anger, for no good will come out of it.

Do not say: "I am your overlord and dictator, and you should, therefore, bow to my commands," as that will corrupt your heart, weaken your faith in religion and create disorder in the state. Should you be elated by power, ever feel in your mind the slightest symptoms of pride and arrogance, then look at the power and majesty of the Divine governance of the Universe over which you have absolutely no control. It will restore the sense of balance to your wayward intelligence and give you the sense of calmness and affability. Beware! Never put yourself against the majesty and grandeur of God and never imitate His omnipotence, for God has brought low every rebel of God and every tyrant of man.

Let your mind respect through your actions the rights of God and the rights of man, and likewise, persuade your companions and relations to do likewise. For, otherwise, you will be doing injustice to yourself and injustice to humanity. Thus both man and God will turn unto your enemies. There is no hearing anywhere for one who makes an enemy of God himself. He will be regarded as one at war with God until he feels contrition and seeks forgiveness. Nothing deprives man of divine blessings or excites divine wrath against him more easily than cruelty. Hence it is that God listens to the voice of the oppressed and waylays the oppressor.

2.4.1 The Common Man

Ali said, 'Maintain justice in administration and impose it on your own self and seek the consent of the people, for the discontent of the masses sterilises the contentment of the privileged few and the discontent of the few loses itself in the contentment of the many. Remember the privileged few will not rally round you in moments of difficulty; they will try to side-track justice, they will ask for more than what they deserve and will show no gratitude for favors done to them. They will feel restive in the face of trials and will offer no regret for their shortcomings. It is the common man who is the strength of the State and Religion. It is he who fights the enemy. So live in close contact with the masses and be mindful of their welfare.

Keep at a distance he who peers into the weaknesses of others. After all, the masses are not free from weaknesses. It is the duty of the ruler to shield them. Do not bring to light that which is hidden, but try to remove those weaknesses which have been brought to light. God is watchful of everything that is hidden from you, and He alone will deal with it. To the best of your ability cover the weaknesses of the public, and God will cover the weaknesses in you which you are anxious to keep away from their eye. Unloose the tangle of mutual hatred between the public and the administration and remove all those causes which may give rise to strained relations between them. Protect yourself from every such act as may not be quite correct for you. Do not make haste in seeking confirmation of tale-telling, for the tale-teller is a deceitful person appearing in the garb of a friend.

2.4.2 The Counsellors

Ali wrote. 'Never take counsel of a miser, for he will vitiate your magnanimity and frighten you of poverty. Do not take counsel of a coward also, for he will cheat you of yourself. Do not take counsel of the greedy, too, for he will instill greed in you and turn you into a tyrant. Miserliness, cowardice and greed deprive man of his trust in God.

The worst of counsellors is he who has served as a counsellor to unjust rulers and shared their crimes. So, never let men who have been companions of tyrants or shared their crimes be your counsellors. You can get better men than these, men gifted with intelligence and foresight, but unpolluted by sin, men who have never aided a tyrant in his tyranny or a criminal in his crime. Such men will never be a burden on you. On the other hand, they will be a source of help and strength to you at all times. They will be friends to you and strangers to your enemies. Choose such men alone for companionship both in privacy and in the public. Even among these, show preference to those who have a habitual regard for truth, however trying to you at times their truth may prove to be, and who offer you no encouragement in the display of tendencies which God does not like his friends to develop.

Keep close to you the upright, and counsellors who are God-fearing, and make clear to them that they are never to flatter you and never to give you credit for any good that you may not have done, for the tolerance of flattery and unhealthy praise stimulates pride in man and makes him arrogant.

Do not treat the good and the bad alike. That will deter the good from doing good, and encourage the bad in their bad pursuits. Recompense every one according one's deserts. Remember that mutual trust and good will between the ruler and the ruled are bred only through benevolence, justice and service. So, cultivate good will amongst the people; for their good will alone will save you from troubles. Your benevolence to them will be repaid by their trust in you, and your ill-treatment by their ill will.

Do not disregard the noble traditions set by our forbearers which have promoted harmony and progress among the people; and do not initiate anything which might minimise their usefulness. The men who had established these noble traditions have had their reward; but responsibility will be yours if they are disturbed. Try always to learn something from the experience of the learned and the wise, and frequently consult them in state matters so that you might maintain the peace and good-will which your predecessors had established in the land.

2.4.3   The Different Classes of People

Ali further wrote. 'Remember that the people are composed of different classes. The progress of one is dependent on the progress of every other; and none can afford to be independent of the other. We have the Army formed of the soldiers of God, we have our civil officers and their establishments, our judiciary, our revenue collectors and our public relation officers. The general public itself consists of Muslims and Zimmis, and among them of merchants and craftsmen, the unemployed and the indigent. God has prescribed for them their several rights, duties and obligations. They are all defined and preserved in the Book of God and in the traditions of his Prophet.

The army, by the grace of God, is like a fortress to the people and lends dignity to the state. It upholds the prestige of the Faith and maintains the peace of the country. Without it the state cannot stand. In its turn, it cannot stand without the support of the state. Our soldiers have proved strong before the enemy because of the privilege God has given them to fight for Him; but they have their material needs to fulfill and have therefore to depend upon the income provided for them from the state revenue. The military and civil population who pay revenue both need the co-operation of others—the judiciary, civil officers and their establishment. The Qazi administers civil and criminal law; the civil officers collect revenue and attend to civil administration with the assistance of their establishment. And then there are the tradesmen and the merchants who add to the revenue of the state. It is they who run the markets and are in a better position than others to discharge social obligations. And then there is the class of the poor and the needy, whose maintenance is an obligation on the other classes. God has given appropriate opportunity of service to one and all; and then there are the rights of all these classes over the administration, which the administrator has to meet with an eye on the good of the entire population, a duty which he cannot fulfill properly unless he takes personal interest in its execution and seeks help from God. Indeed, it is obligatory on him to impose this duty on himself and to bear with patience the inconveniences and difficulties incidental to his task.

2.4.4   The Army

Ali wrote. 'Be particularly mindful of the welfare of those in the army who, in your opinion, are staunchly faithful to their God and Prophet and loyal to their chief, and who in the hour of passion can restrain themselves and listen coolly to sensible remonstrance, and who can succour the weak and smite the strong, whom violent provocation will not throw into violent temper and who will not falter at any stage.

Keep yourself in close contact with the families of established reputation and integrity with a glorious past, and draw to yourself men brave and upright in character, generous and benevolent in disposition, for such are the salt of society.

Care for them with the tenderness with which you care for your children, and do not talk before them of any good that you might have done to them, nor disregard any expression of affection which they show in return, for such conduct inspires loyalty, devotion and goodwill. Attend to every little of their want, not resting content with what general help that you might have given to them, for sometimes, timely attention to a little want of theirs brings them immense relief. Surely these people will not forget you in your own hour of need.

It behooves you to select for your Commander-in-chief one who imposes on himself as a duty the task of rendering help to his men, and who can excel in kindness every other officer who has to attend to the needs of the men under him, and look after their families when they are away from their homes; so much so, that the entire army should feel united in their joys and in their sorrows. The unity of purpose will give them added strength against the enemy. Continue to maintain a kindly attitude towards them so that they might feel attached to you. The fact is that the real happiness of the administrators and their most pleasant comfort lies in establishing justice in the state and maintaining affectionate relations with the people. Their sincerity of feeling is expressed in the love and regard they show to you, on whom alone depends the safety of the administrators.

Your advices to the army will be of no avail, unless and until you show affection to both men and officers, in order that they might not regard the Government as an oppressive burden or contribute to its downfall.

Continue to satisfy their needs and praise them over and over again for what services they have rendered. Such an attitude, God willing, will inspire the brave to braver actions and induce the timid to deeds of bravery.

Try to enter into the feelings of others and do not foist the mistake of one over another and do not grudge dispensing appropriate rewards. See to it you do not show favours to one who has done nothing, but merely counts on his family position; and do not withhold proper rewards from one who has done great deeds simply because he holds a low position in life.

2.4.5 The Real Guidance

Ali advised. 'Turn to God and to His prophet for guidance whenever you feel uncertain as to what you have to do. There is the commandment of God delivered to those people who He wishes to guide aright: "O people of the Faith! Obey God and obey His prophet and those from among you who hold authority over you." And refer to God and His prophet whenever there is difference of opinion among you. To turn to God is in reality to consult the Book of God; and to turn to the prophet is to follow his universally accepted traditions.

2.4.6   Chief Justice

Ali advised. 'Select for you a chief judge from your people who is by far the best among them—one who is not obsessed with domestic worries, one who cannot be intimidated, one who does not err too often, one who does not turn back from the right path once he finds it, one who is not self-centred or avaricious, one who will not decide before knowing full facts, one who will weigh with care every attendant doubt and pronounce a clear verdict after taking everything into full consideration, one who will not grow restive over the arguments of advocates and who will examine with patience every new disclosure of fact and who will be strictly impartial in his decision, one whom flattery cannot mislead or one who does not exult over his position. But it is not easy to find such men.

Once you have selected the right man for the office, pay him handsomely enough, to let him live in comfort and in keeping with his position, enough to keep him above temptations. Give him a position in your court so high none can even dream of coveting it and so high that neither back-biting nor intrigue can touch him.

2.4.7   Subordinate Judiciary

Ali further advised. 'Beware! The utmost carefulness is to be exercised in his selection, for it is this high office which adventurous self-seekers aspire to secure and exploit in their selfish interests. After the selection of your chief judge, give careful consideration to the selection of other officers. Confirm them in their appointments after approved apprenticeship and probation. Never select men for responsible posts either out of any regard for personal connections or under any influence, for that might lead to injustice and corruption.

Of these, select for higher posts men of experience, men firm in faith and belonging to good families. Such men will not fall an easy prey to temptations and will discharge their duties with an eye on the abiding good of others. Increase their salaries to give them a contented life. A contented living is a help to self-purification. They will not feel the urge to tax the earnings of their subordinates for their own upkeep. They will then have no excuse either to go against your instructions or misappropriate state funds. Keep watch over them without their knowledge, loyal and upright men. Perchance they may develop true honesty and true concern for the public welfare. But whenever any of them is accused of dishonesty, and the guilt is confirmed by the report of your secret service, then regard this as a sufficient to convict him. Let the punishment be corporal and let that be dealt in the public at an appointed place of degradation.

2.4.8   Revenue Administration

Ali warned. 'Great care is to be exercised in revenue administration, to ensure the prosperity of those who pay the revenue to the state; for on their prosperity depends the prosperity of others, particularly the prosperity of the masses. Indeed, the state exists on its revenue. You should regard the proper upkeep of the land in cultivation as of greater importance than the collection of revenue, for revenue cannot be derived except by making the land productive. He who demands revenue without helping the cultivator to improve his land inflicts unmerited hardship on the cultivator and ruins the State. The rule of such a person does not last long. If the cultivators ask for reduction of their land tax for having suffered from epidemics or drought or excess of rains or the barrenness of the soil or floods damaging to their barrenness of the soil or foods damaging to their crops, then, reduce the tax accordingly, so that their condition might improve. Do not mind the loss of revenue on that account for that will return to you one day manifold in the hour of greater prosperity of the land and enable you to improve the condition of your towns and to raise the prestige of your state. You will be the object of universal praise. The people will believe in your sense of justice. The confidence which they will place in you in consequence will prove your strength, as they will be found ready to share your burdens.

You may settle down on the land any number of people, but discontent will overtake them if the land is not improved. The cause of the cultivator's ruin is the rulers who are bent feverishly on accumulating wealth at all costs, out of the fear that their rule might not last long. Such are the people who do not learn from examples or precedents.

2.4.9   Clerical Establishment

Ali warned. 'Keep an eye on your establishment and your scribes; and select the best among them for your confidential correspondence such among these as possess high character and deserve your full confidence, men who may not exploit their privileged position to go against you and who may not grow neglectful of their duties and who in the drafting of treaties may not succumb to external temptation and harm your interests, or fail to render you proper assistance and to save you from trouble, and who in carrying out their duties can realize their serious responsibilities, for he who does not realise his own responsibilities can hardly appraise the reprehensibilities of others. Do not select men for such work merely on the strength of your first impressions of your affection or good faith; for as a matter of fact, the pretensions of a good many who are really devoid of honesty and good breeding may cheat even the intelligence of rulers. Selection should be made after due probation which should be the test of righteousness. In making direct appointments from people, see to it that those selected possess influence with the people and who enjoy the reputation of being honest; for such selection is agreeable to God and the ruler. For every department of administration, let there be a head, to whom no trying task might cause worry and no pressure of work annoy.

And remember that every weakness of any one among your establishment and scribe which you may overlook will be written down against you in your scroll of deeds.

2.4.10 Trade and Industry

Ali advised. 'Adopt useful schemes placed before those engaged in trade and industry and help them with wise counsels. Some of them live in towns, and some move from place to place with their wares and tools and earn their living by manual labor. Trade and Industry are sources of profit to the State. While the general public is not inclined to bear the strain, those engaged in these professions take the trouble to collect commodities from far and near, from land and from across the sea, and from mountains and forests and naturally derive benefits.

It is this class of peace-loving people from whom no disturbance need be feared. They love peace and order; indeed, they are incapable of creating disorder. Visit every part of the country and establish personal contact with this class, and inquire into their condition. But bear in mind that a good many of them are intensely greedy and are inured to bad dealings. They hoard grain and try to sell it at a high price; and this is most harmful to the public. It is a blot on the name of the ruler not to fight this evil. Prevent them from hoarding; for the Prophet of God, Peace be upon him, had prohibited it. And see to it that trade is carried on with the utmost ease, that the scales are evenly held and that prices are so fixed that neither the seller nor the buyer is put to a loss. And if in spite of your warning, should anyone go against your commands and commit the crime of hoarding, then deal him appropriately with severe punishment.

2.4.11 The Poor

Ali admonished. 'Beware! Fear God when dealing with the problem of the poor who have none to patronise, who are forlorn, indigent and helpless and are greatly torn in mind—victims of the vicissitudes of Time. Among them there are some who do not question their lot in life, who not withstanding their misery, do not go about begging. For God's sake, safeguard their rights; for on you rests the responsibility of protection. Assign for their uplift a portion of the state exchequer (Baitul-mal), wherever they may be, whether close at hand or far away from you. The rights of the two should be equal in your eye. Do not let any preoccupation slip them from your mind; for no excuse whatsoever for the disregard of their rights will be acceptable to God. Do not treat their interests as of less importance than your own, and never keep them outside the purview of your important considerations, and mark the persons who look down upon them and of whose conditions they keep you in ignorance.

Select from among your officers such men as are meek and God-fearing who can keep you properly informed of the condition of the poor. Make such provision for these poor people as shall not oblige you to offer an excuse before God on the day of judgment; for it is this section of the people more than any other which deserves benevolent treatment. Seek your reward from God by giving to each of them what is due to him and enjoin on yourself as a sacred duty the task of meeting the needs of such aged among them as have no independent means of livelihood and are averse to seek alms. And it is the discharge of this duty that usually proves very try-ing for ruler, but is very welcome to societies which are gifted with foresight. It is only such societies or nations who truly carry out with equanimity their covenant with God to discharge their duty to the poor.

2.4.12 Open Conferences

Ali advised. 'Meet the oppressed and the lowly periodically in an open conference and, conscious of the divine presence there, have a heart-to-heart talk with them, and let none from your armed guard or civil officers or members of the police or the Intelligence Department be by your side, so that the representatives of the poor might state their grievances fearlessly and without reserve. For I have the Prophet of God saying that no nation or society will occupy a high position in which the strong do not discharge their duty to the weak. Bear with composure any strong language which they may use, and do not get annoyed if they cannot state their case lucidly; even so, God will open you his door of blessings and rewards. Whatever you can give to them, give it ungrudgingly, and whatever you cannot afford to give, make that clear to them in utmost sincerity.

There are certain things which call for prompt action. Accept the recommenda-tions made by your officers for the redress of the grievances of the clerical staff. See to it that petitions or applications that are submitted for your consideration are brought to your notice the very day they are submitted, however much your officers might try to intercede them. Dispose of the day's work that very day, for the coming day will bring with it its own tasks.

2.4.13 Communion with God

Ali commanded. 'And do not forget to set apart the best of your time for communion with God, although every moment of yours is for Him only, provided it is spent sincerely in the service of your people. The special time that you give to prayer in the strict religious sense is to be devoted to the performances of the prescribed daily prayers. Keep yourself engaged in these prayers both in the day and in the night, and to gain perfect communion, do not, as far as possible, let your prayers grow tire-some. And when you lead in congregational prayer, do not let your prayer be so lengthy as to cause discomfort to the congregation or raise in them the feeling of dislike for it or liquidate its effect, for in the congregation there may be invalids and also those who have to attend pressing affairs of their own.

When I had asked of the Prophet of God on receiving an order to proceed to Yemen, how I should lead the people there in prayer, he said perform your prayers even as the weakest among you would do; and set an example of consideration to the faithful.

2.4.14 Aloofness or Isolation

Ali reprimanded. 'Along side the observance of all that I have said above bear one thing in mind. Never for any length of time keep yourself aloof from the people, for to do so is to keep oneself ignorant of their affairs. It develops in the ruler a wrong perspective and renders him unable to distinguish between what is important and what is not, between right and wrong, and between truth and falsehood. The ruler is after all human; and he cannot form a correct view of anything which is out of sight. There is no distinctive sign attached to truth which may enable one to distinguish between the different varieties of truth and falsehood. The fact is that you must be one of two things. Either you are just or unjust. If you are just, then you will not keep yourself away from the people, but will listen to them and meet their requirements. On the other hand, it you are unjust, the people themselves will keep way from you. What virtue is there in your keeping aloof? At all events aloofness is not desirable especially when it is your duty to attend to the needs of the people. Complaints of oppression by your officers or petitions for justice should not prove irksome to you.

Make it clear to yourself that those immediately about and around you will like to exploit their position to covet what belongs to others and commit acts of injustice. Suppress such a tendency in them. Make a rule of your conduct never to give even a small piece of land to any of your relations. That will prevent them from causing harm to the interests of others and save you from courting the disapprobation of God and Man.

Deal justice squarely regardless of whether one is a relation or not. If any of your relations or companions violates the law, mete out the punishment prescribed by law, however painful it might be to you personally, for it will be all to the good of the State. If at any time people suspect that you have been unjust to them in any respect disclose your mind to them and remove their suspicions. In this way, your mind will get attuned to the sense of justice and people will begin to love you. It will also fulfil your wish that you should enjoy their confidence.

2.4.15 Peace and Treaties

Ali warned. 'Bear in mind that you do not throw away the offer of peace which your enemy may himself make. Accept it, for that will please God. Peace is a source of comfort to the army; it reduces your worries and promotes order in the State. But beware! Be on your guard when the peace is signed, for certain types of enemies propose terms of peace just to lull you into a sense of security only to attack you again when you are off your guard. So you should exercise the utmost vigilance on your part, and place no undue faith in their protestations. But, if under the peace treaty you have accepted any obligations, discharge those obligations scrupulously. It is a trust and must be faithfully upheld and whenever you have promised anything, keep it with all the strength that you command, for whatever differences of opinion might exist on other matters, there is nothing as noble as the fulfilment of a promise. This is recognised even among non-Muslims, for they know the dire consequences which follow from the breaking of covenants. So never make excuses in discharging your responsibilities and never break a promise, nor cheat your enemy. For breach of promise is an act against God, and none except the positively wicked acts against God.

Indeed, divine promises are a blessing spread over all mankind. The promise of God is a refuge sought after even by the most powerful on earth; for there is no risk of being cheated. So, do not make any promise from which you may afterwards offer excuses to retract; nor do you go back upon what you have confirmed to abide by; nor do you break it, however galling it may at first prove to be. For it is far better to wait in patience for wholesome results to follow than to break it out of any apprehensions.

Be warned! Abstain from shedding blood without a valid cause. There is nothing more harmful than this which brings about one's ruin. The blood that is willfully shed shortens the life of a state. On the day of judgment it is this crime for which one will have to answer first. So, beware! Do not wish to build the strength of your state on blood; for it is this blood which ultimately weakens the state and passes it on to other hands. Before me and my God, no excuse for willful killing can be entertained.

Murder is a crime which is punishable by death. If on any accord the corporal punishment dealt by the state for any lesser crime results in the death of the guilty, let not the prestige of the stage stand in any way of the deceased relations claiming blood money.

2.4.16 Final Instructions

Ali gave his final instructions. 'Do not make haste to do a thing before its time, nor put it off when the right moment arrives. Do not insist on doing a wrong thing, nor show slackness in rectifying a wrong thing. Perform everything in its proper time, and let everything occupy its proper place. When the people as a whole agree upon a thing, do not impose your own view on them and do not neglect to discharge the responsibility that rests on you in consequence. For the eyes of the people will be on you and you are answerable for whatever you do to them. The slightest dereliction of duty will bring its own retribution. Keep your anger under control and keep your hands and tongue in check. Whenever you fall into anger, try to restrain yourself or else you will simply increase your worries.

It is imperative on you to study carefully the principles which have inspired just and good rulers who have gone before you. Give close thought to the example of our prophet (peace be upon him), his traditions, and the commandments of the Book of God and whatever you might have assimilated from my own way of dealing with things. Endeavour to the best of your ability to carry out the instructions which I have given you here and which you have solemnly undertaken to follow. By means of this order, I enjoin on you not to succumb to the prompting of your own heart or to turn away from the discharge of duties entrusted to you.

I seek the refuge of the might of the Almighty and of His limitless sphere of blessings, and invite you to pray with me that He may give us together the grace willingly to surrender our will to His will, and to enable us to acquit ourselves before Him and His creation; so that mankind might cherish our memory and our work survive. I seek of God the culmination of his blessings and pray that He may grant you and me His grace and the honour of martyrdom in His cause. Verily, we have to return to Him. I invoke His blessings on the Prophet of God and his pure progeny.

This commandment by Ali Ibne Abi Talib, was distributed by General Abbasi, a Governor of Sindh province some 20 years ago as a manual for all officers to follow. It has been widely circulated by many other rulers over centuries for ideal governance observing the most humane code of conduct, and has been translated into innumerable languages for nations and people to follow. If only people could follow his advice! If they did, the world could be highly moral, ethical and righteous. Justice, equity, fair play and devolution of responsibilities, as indeed matching rights, would bring about peace and harmony in all groups that form a part of a community.

One salient attribute that the physician must possess is the quality of Leadership and Team work. It was duly identified in the famous GMC document called *Tomorrow's Doctor?*

In medicine, those of us who hold an administrative position or enjoy a status in their field as an expert or an educator, should follow the mandates given by Imam Ali. It would bring about ethical practices back into our profession, which somehow have been put aside for many reasons.

One does not have to go far looking for malpractices in NHS and abuse of authority. Ask any person in employment in the NHS today and you would get the same answer: The managers have made their life miserable to meet their targets, thus compromising the services beyond description.

The Mid Stafford Hospital disaster of 2013 is one glaring example of neglect of patient care, abuse of authority, and the deprivation or withholding of treatment to the elderly, the aged and the disabled. It has since gone into administration.

In many other less developed nations, a physician holding a position of authority is no less than a tyrant and a dictator in his own right. He not only controls the destiny of his subordinates and twists and turns their arms to his wishes, but often enough treats them as slaves.

Inter-professional ethics is another issue, which is all but absent in most countries. The Hippocratic Oath defines the basic parameters, such as mutual respect between colleagues as well as respect and honour to be accorded to not just the teacher but even his wards.

It is seldom, if ever, seen in the medical profession, just as in other professions. Petty jealousies, backbiting, bickering and discreet and sometimes hate campaigns are the order of the day these days. The new inventions of electronic media, such as e-mails, texts, Twitter, Facebook and LinkedIn are meant for exchanging urgent messages, maintaining communication and for general social exchange, but they are more often than not used for defamation, allegations, and hate mail. It is pathetic and disgraceful, but it goes on unabated.

Sycophancy is an art mastered by physicians in the developing countries, where even a Professor and chair has to apply this Machiavellian technique to please the bosses and torture his subordinates.

Insecurity breeds greed which breeds tyranny. So we see the likes of that devil called Saddam or his cohort Ghadafi and their cronies in many other countries become historical tyrants, only to be disgraced by their sycophant mobs over time.

Professional rivalry in medicine is inevitable and somewhat acceptable. It may yield fruitful results through competition, but jealousy and abusive practices are also not uncommon. I have seen, in some places, colleagues not even talking to each other let alone sitting down for an exchange of ideas or any formative opinions for their mutual benefit. It is a practice that must be condemned. We should all learn lessons on human attitude, behaviour, and interpersonal ethics from this superb epistle of Imam Ali.

References

1. Quran Al Shua'ara 26:80
2. Health and dharma in hindu ayurvedic medicine. p 120. Ayurvedic- Carakashimta. www.nccm. nih.gov/health/Ayuervedic/introduction.htm
3. Desai P (1988) Medical ethics in India. J Med Philos 244–246

Nomenclature and Descriptive Analysis

3.1 Sources and Pillars of Medical Ethics

From times gone by, medicine and medical ethics have remained an integral part of the profession. Islamic medicine is full of tales of the contributions made by eminent *hukama* of the golden Islamic period between the sixth and fourteenth centuries AD: men like Al Razi, writing his *Tibee Ruhani*; or *Adab al Tibb* by Al-RUhawai, a Muslim convert from Christianity; or the illustrious *Ferdous al Hikama* by Al Tabari and *Kamale al Sanaat al Tibia* by Ahwazi; not to mention the famous works of Ibne Sina in *Al Qanoon fil Tibb*. These giants of medicine have left their lasting impressions on the human mind and history.

Medical ethics has applied many of these principles in its daily practice, albeit without exactly defining the terms of reference per se. It was the combination of two modern philosophers, Beauchamp and Childress, who in 1994 defined the following four normative principles of medical ethics that are now applied in medical practice across the world. In fact, all decisions by ethics committees must be made under these guidelines when a serious question of an ethical nature arises. These principles are

Autonomy
Beneficence
Nonmalfeasance
Justice

3.1.1 Autonomy

Autonomy basically means an individual's rights to make his own decisions in all matters of life and death, with full and detailed information, including benefits and risks involved, or the consequences of refusing to accept advice given by an expert. It primarily focuses on the human rights of an individual, dignity and respect, freedom of thought and expression, self-determination, self-esteem and most of all choosing a path for one's best interest free of pressure, coercion or fear, with

S.H. Zaidi, *Ethics in Medicine*,
DOI 10.1007/978-3-319-01044-1_3, © Springer International Publishing Switzerland 2014

individual dignity, pride and honour. Autonomy, however, has many forms to offer. It is deeply influenced by the culture, traditions, norms, and values of a given society. A socio-anthropological study of autonomy would prove beyond the shadow of a doubt that when it comes to autonomy, the old dictum of 'different horses for different courses' is most appropriately suitable. In Western societies, erosion of authority has seeped through the culture so deeply that not just the trust in the governance but indeed the very fabric of the society is jeopardized by a lack of trust and mutual faith. The family units are nonexistent and therefore everyone is to himself; thus no one has any form of authority to make a decision for a family member. In many ways it upholds the fundamental rule of free decision making, but many times even a free soul needs guidance and advice, which regrettably is not available in the hour of need. Frankly, when you become as self-centered as many members of Western societies have, you are bound to be isolated. It is a common observation in any medical practice in the UK, as indeed in the USA, that many young boys and girls, particularly of a certain age, tend to drift away from the norms into the abyss of depression, solitude, and misery. Drugs and similar temporary measures become their tools of respite and rescue, only to make matters worse. In fact, the teenage pregnancy rate (30 per 1,000 population in 2013 in the UK) and so-called 'love children' seen so often in British society today, are the outcome of excessive freedom, autonomy beyond measure, and total lack of authority. Binge drinking make it supremely worse, and in 2012 the Cameron government was left with no option but to resort to stricter controls and, as he said in one of his forceful speeches, 'revival of moral values' to overcome the growing menace of a degenerating society.

On the other hand, the excessive abuse of power and total deprivation of autonomy, fundamental rules, and personal decision making are commonly observed in Asia, the Middle East, and Africa. In the Muslim countries of these regions, the head of the tribe, clan, or family has total authority. No one can challenge his authority or his decisions. It is taken for granted that whatever the head decides is in the best interest of the ward. This particularly manifests itself in relationship to women. Many such societies are still so primitive that the female population has no control over their destiny. The might in such societies is always right. We see it in daily life as indeed in all matters of governance in many countries and nations, from Morocco, to Libya, to Egypt, and across the Arabian Peninsula to Indo-Pakistan and Far Eastern countries like Indonesia. Malaysia is somewhat different, where women have acquired a major say in decision making.

Untold stories of the misery sustained by women in Pakistani rural areas are on daily display on national television. A recent illustration was seen when a law-making parliamentarian ruthlessly slapped a poor polling officer for an unknown reason. It was recorded and displayed by the media ad nauseam. Daily stories of acid splashing on the faces of women by their menfolk was duly documented by Shirmeen Chinoy, a Karachi *grammar* school girl in an Oscar award-winning documentary in 2012. Has it changed the situation? Regrettably not. The untold stories of hapless women continue unabated and wreak havoc in rural Pakistani society. Worse is the situation in Balochistan, where women are literally kept in the enclosure of their homes forever. The renowned historian Mubarak Ali once analysed the

Pajerro culture in Pakistan, describing a psychological phenomenon of a show of authority and a way of keeping women out of sight behind the dark windows of those mysterious vehicles while having a jolly good view of other women on the roads and shopping centres in Clifton or Defence Society.

So, what is the answer? It is simple. The Prophet set up an excellent example in his life when he granted a special position to women in his sermons and displayed similar respect in his practical life. One fine example is that whenever Syeda Fatima ut Zahra visited her father, the Prophet would stand up to greet her and would wait for her to take a seat before settling down himself. The narrators have informed us that his respect for Janab Khatidja tul Kubra was legendary and lasted until he passed away, as it is said that he could not forget her devotion, help and association with him despite the fact that she had passed away many years earlier.

The denial of human rights is currently a major issue around the world. Prophet Muhammad, in his message at *Hajj al Wida*, liberated mankind through his lasting message. He forgave the blood money, or *qisas*, for his dead and gave full liberty to the slaves who had suffered for generations at the hands of mighty Quraish. He gave equal rights to all mankind, irrespective of colour, creed, race or kinship. His message was memorised by thousands attending his sermon, and yet much of it was soon forgotten. The sad story of Islam is that *malookiat* took over the reins of control not a couple of decades after the great messenger passed away. The Byzantine influence, compounded with the glory of *Khusaraus* of Iran, soon infiltrated the polluted palaces of Banu Umayyah. Syria became the controller of the destiny of Muslim Ummah, and the last of the founding pillars of faith was hastily demolished in the *Masjide Koofa*. Hussain saw it all, imbibed it thoroughly, and studied the events with the *Aql o Idrak* that only an Imam possesses. He planned with precision. He had seen his ancestors succumb to the treachery of Quraish and its dominant clan. He had seen his father appeal for support and then yell at the Koofis, cursing them for failing to rise against the evil forces. He also had seen peace with a cunning enemy that his pious brother made which did not work. So he decided to give an electric jolt to the Muslim psyche. He embarked on his mission with the statement that is, in fact, the fundamental brick of Islam, that is, "*Amr bil maroof wa nahi anal munkar.*" He sacrificed everything to shock the Muslim mind to wake up to the bitter realty that what Prophet SA came to bestow was all but exterminated by the evil forces of Syria. He upheld truth, justice, honesty, integrity, human rights, denial of might, and establishment of the rights of the poor and downtrodden, the orphans and widows, just as his father and grandfather had done. What better example is there than his, a perfect model for a Muslim to study and duly apply to the basic principles of autonomy. The word *autonomy* is basically a composite term. Its elements are consent, confidentiality, anonymity, truth, faith and trust.

Consent has many colours. It can be an 'express consent' as seen in day to day clinical consent. It does not need to be documented per se; verbal consent is good enough. For instance, for a nurse to give an injection or for a doctor to do a simple nasoendoscopy, all that is needed is that the procedure and its risks as well as benefits may be clearly explained to the patient in a few simple words. But the more usual form of consent is called 'informed consent.' In fact, the word *consent* first came in use after the Nuremberg trials and formation of the Nuremberg code in 1948. The phrase 'informed consent' was first formally used by a judge in California in 1957. Now it has become the byword for every form of medical or surgical intervention. Any clinician would know that it is now mandatory to explain in simple, non-technical terms the procedure we plan to carry out, its merits, indications, and benefits and a detailed list of all possible risks. On 17 April 2012 the British authorities overruled the time honoured Bolam principle. The Bolam principle said that any complication arising from a procedure by a trained and authorized person may be an acceptable risk if a court finds that, in a similar situation, another reasonably qualified person would not have mentioned the risks prior to consent for an intervention or a procedure such as surgery. Furthermore, a physician is within his rights to invoke a 'therapeutic privilege' by either withholding or not mentioning a risk of a procedure if he believes that the information thus given may cause more harm than benefit to the patient.

Now the practice has been modified so that all major and minor complications must be mentioned and duly recorded in the notes and the consent form, before a procedure is carried out. Some quarters have already raised their concerns in this respect as it may further deter even the most genuine patients from undergoing elective surgery. Anyone familiar with the NHS may have noticed that surgeons are becoming more and more conservative. It has often been described as defensive medicine. The quality of surgical practice has significantly dropped as the society becomes more litigious by the day.

The information provided to patients may be not just verbal, but all forms of literature support, such as brochures, leaflets and pamphlets, must also be provided, and duly recorded in the case notes.

In case of an adult, it is not such an issue, as most Western societies have been trained to do so, right from their early ages. They are educated, at least to a certain extent if not highly educated, but they can demand all the information they need to know and the physician is obliged to fill them up with the required details.

'Informed consent' has been defined by the pundits as an autonomous action taken by a patient, thereby authorizing a physician to initiate a plan of action for medical care. It is considered to be only valid if the patient has substantial understanding of the problem and the intervention or remedy offered by the physician including the merits and the risks involved. It is also valid only if the patient is competent enough to understand the implications of the disease and its remedy offered. Finally, the informed consent is considered valid only if the criteria, namely the substantial independence and freedom to make such a decision, is available to the patient.

As mentioned previously, an 'express consent' is a common form of consent in which it is presumed that the patient is fully cognizant of the procedure or the effects of such a procedure and agrees to it verbally, or in a written form. It basically means that the consent is implied by the actions of the patient in a given situation. The old saying that sometimes actions are more meaningful than words is a good illustration here. If the patient does not offer a resistance to a given proposal or an action, and volunteers to an examination, etc., then a written or a verbal consent may not be required, and it may by implied from the action alone that he has consented to a given procedure. The best example would be a routine examination, after the physician has noted down the history and would now proceed with a routine examination or order a blood or urine test, etc.

Sometimes the above forms of consent may not be practical, such as in an emergency situation. Here the term used is the 'general consent.' It basically implies that because of the situation, the patient has given a general consent to proceed with the emergency treatment. The patient is required to sign an agreement form, at the time of reporting to an emergency room, to allow the medics to proceed with the emergency evaluation and treatment, but it does not replace an informed consent per se, which must be obtained as the situation improves and further intervention may be required.

In emergency medicine, another form of consent that is employed is called the 'specific consent.' It is required when a more invasive procedure may be required and is akin to an informed consent.

And sometimes the situation may demand that an 'emergency consent' may be employed, when the patient's situation is so critical that there may not be time enough to proceed with the formalities. Such are specifically the situations in which the physician must observe the oath of ethics or professional oath that he had taken when entering the profession. He must do all he can with beneficence and nonmalfeasance to save a life.

'Deferred consent' is sometimes employed as a part of research on an unresponsive patient to proceed with the treatment or a procedure until a formal consent is obtained by the patient on recovery or relief or by the next of kin.

Informed consent, in fact, is the type of consent that must be obtained in all definitive treatments. Its fundamental element are

1. A fair and detailed explanation of the procedure, or intervention, and their objectives, including identification of any subsequent process or procedures that may be required.
2. Detailed description of the objectives, risks and possible complications ensuing there from, as indeed the benefits the procedure will bring to the patent.
3. The alternative treatments that may resolve the patient's problem should also be discussed.
4. Asking the patient to answer if any queries or doubts need to be answered or resolved.
5. The choice to withdraw or cancel a proposed treatment or procedure must remain with the patient, totally free of an authoritative enforcement.

Since informed consent is akin to a contract between the patient and the server, all papers must be duly signed, dated and a copy given to the patient. No alteration must be made in this legal document, once signed and sealed. An overwriting or correction and rewriting, etc., must be avoided. If the procedure needs any addition or subtraction, always use a fresh form and duly cancel the previous one, making a note of it in the case notes while retaining a copy of the cancelled document in the file.

As mentioned before, the phrase 'informed consent' was first ever used by Justice Bray in a California court of appeals in 1957, when Judge Bray wrote, "A physician violates his duty to his patient and subjects himself to a liability if he holds back any facts which are necessary to form the basis of an intelligent consent by the a patient to the proposed treatment; discussing the elements of risk a certain discretion must be employed consistent with the full disclosure of facts necessary for an informed consent."

Rapid advancement in knowledge and technology have brought about many changes in keeping the patient fully informed as indeed in the safety of keeping the records, etc. But more importantly, the time-honoured, paternalistic approach of health care is gone and done with. Both the physician and the patient are each now a party to an arrangement reached with joint consensus. Each party must respect the other party's views.

Another radical change that has taken place is that while the family physician was almost like a family member, who knew the patient and his family just like his own and enjoyed trust and confidence of the family, the specialist is almost like a third party. He is usually not so close to the family, indeed often enough a total stranger. Therefore, the bond of love, faith and mutual trust is lacking in modern-day specialist practice. Suspicion breeds fear and a lack of confidence. Therefore, an informed consent has now become all the more relevant and has acquired an unprecedented place in current medical practice in most developed and many emerging economies.

The following information should be provided to the patient:

The diagnosis of the problem

The nature and purpose of the treatment offered

Risks and benefits of the treatment offered

Alternate treatments available

Prognosis of the proposed treatment

Answers to any queries raised by the patient

If not sure, allow the patent time to think it over and return at later date.

If unconvinced, a second opinion may be sought.

The list of items may be exhaustive, but suffice it to say that caution is always a better option in all matters requiring informed consent, particularly in the current environment of litigations and solicitations for damages, etc., so often advertised by the stake holders in the media.

There are many grey zones in the matter of informed consent, for instance, in a case of a mentally disabled person of a certain age. In a society in which such a person is under the umbrella of altruist family members, a surrogate consent may be acceptable; but in the British society it is a major issue, as many such persons live under the sheltered environment of a third-party caregiver. Here the issues must be discussed with the social agencies, the caregivers and in full cognizance of the GP, the local authority and the next of kin. More details can be found in the relevant literature through the net.

In the case of children and minors, once again it is not much of a problem in the developing countries, in which family ties are strong and binding and consent will be obtained from the parents. In the developed nations, where many a social problems and demise of the family units as indeed horrible stories of pedophilia, etc., continue to circulate, matters must be dealt with great caution. The informed consent in children living with the foster parents and care homes is a very special matter too. All social and legal agencies as well as GP must be taken into confidence prior to any intervention. Guidelines are published by the health authorities in this respect from time, which must be followed. Currently the age for giving consent personally in the UK is 16 years.

Research ethics is a subject in its own right and will be dealt with accordingly. In experimental research or clinical trials, informed consent becomes highly important and will be discussed in that chapter.

Screening for many potential illnesses has become a routine in many countries. It is debatable if screening for such conditions as Alzeihmers and colonic cancer are currently in progress in the UK; they may also require informed consent before such a survey is carried out. Screening may sometimes lead to an unwanted rise in health insurance cost, applicable to many countries where this practice exists. It may also jeopardize patients' interests for the same reasons. Besides, misinterpretation or misunderstanding of the results can lead to confusion, psychological stress and worse, i.e., stigmatization. It is therefore essential that any form of a screening programme should be cleared by the national ethics committee, and informed consent should automatically become part of the recruitment process.

The mental disability act in practice in the UK duly highlights the fact that one may not go by hearsay or the appearance of a patient to determine if or not he is mentally deficient. Simple interviews should help determine the real picture. If he is really deficient and there is no surrogate, a responsible caretaker or preferably a relative, then the physician is obliged to make decision, keeping the best interest of the patient in the forefront.

At the 2013 IMI Kufa conference, much time was dedicated to the subject of ethics. Sarwat Hussain, an eloquent speaker, gave an inspirational lecture on various forms of consent, as applicable to contemporary medicine. He also highlighted the significance of surrogate consent in some cultures. He did not agree, though, with other speakers that religion and ethics are separate entities. Sarwat believes that religion is ethics and its fundamental principles, as dictated in Quran and practiced by the Prophet, is nothing but ethics. Therefore, in his view one cannot differentiate between religion and ethics; indeed they are part and parcel of the whole principle of morality. In other words, the whole (religion) encompasses all parts (principles of ethics) and is better than the sum of the parts.

Presumed Consent
This is the latest addition to the field of consents. It is currently being discussed and debated in many circles, including the British legislative authorities and the hierarchy in the United States. It is anticipated that within the legislation of presumed consent, the default rule would be an agreement and an implied consent for organ donation, which is where this particular form of consent will be most applicable. Unless a person opts out by writing down in his directive or a living will, it will be presumed that his organs can be harvested. It also means that all such measures that may be necessary to procure the organs by preserving them at the end of life,

including delaying death and prolonging the dying process, interventions, medications and life support systems, will be legal and ethical under the heading of presumed consent. There are many reasons for a Muslim to object to such a practice. Besides, one must remember that when a life-support system is introduced for delaying death in order to procure the organs, without consent, the interests change from the care and service of the dying to that of the third party, i.e., the organ-harvesting team or the ultimate beneficiary. It is therefore necessary that first-person consent be obtained.

Islam strongly believes in avoiding any delay in the final settlement of the body. Any undue delay, even for organ harvesting, may create unpleasantness in the family, friends the *Bradari* or the tribe. History informs us that when the Prophet breathed his last Ali and his brothers stayed with the dead body and performed the religious rituals, buried the Prophet and offered *Namaze janaza*, while the Ummah and its prominent Quraish leaders, as indeed the Ansars, left to sort out Khilafat.

It must, therefore, be said that delaying burial may not be the practice of the followers of Ahle Beit, and of the majority of Muslims, thereby challenging the authenticity of presumed consent. As has been discussed elsewhere in this book, ethics is culture-sensitive, and one can't have 'one size fits all' in such delicate matters as death and organ harvesting.

Trust and Confidentiality

These two elements are an integral part of autonomy. From times immemorial a covenant of trust existed between the doctor and his patient. Many times a patient would confide more in his doctor than even to his spouse, and the doctor felt it a religious duty to safeguard that secret. That trust was best displayed in the story told by a renowned historian about Dr. Mistry, a Parsi gentleman. Dr. Mistry was Quiade Azam Mohammad Ali Jinnah, the personal physician of the founder of Pakistan. When Mr. Jinnah was informed that his cough and ill health were caused by a fatal illness, allegedly tuberculosis, which in 1945–1946 was an incurable condition, it is said that Mr. Jinnah asked the good doctor to keep it secret; and the doctor dutifully obliged. Such are the doctors that are legendary. But there are not many, dare I say!

Ever since the demise of paternalistic professional care and the loss of family units, the doctor-patient relationship has changed permanently. Certain parameters must therefore be defined and practiced in all matters of health care.

Patients have a right to confidentiality and expect that their profiles, as indeed their data, must be safeguarded. In fact, due to a mishap in the NHS not long ago, wherein patient data was lost, stolen or hacked into, most NHS trusts have encrypted the USBs for computers in the hospitals and practices.

It is not without a reason that the patients demand secrecy of their data and medical conditions. The whole world has become greedy, and many family members may want to know when the rich uncle would die, so they could share his wealth by false pretences or even malicious acts.

Sometimes the data may have to be shared. In such situations the physician must seek written approval of the patient to release the information. It is always advised that data should be anonymous in all those situations in which unidentifiable data would be acceptable to serve the purpose. Data may be transferred within the trust even across the NHS, but all staff members are honour-bound to keep the disclosure to a third part at abeyance and, if at all required, then must keep it anonymous.

Patients may demand to know of the investigative results. They have a right to do so. The physician must oblige and note it down in the case notes. If the patient wants to know about the disease, its management and prognosis, etc., a physician must respect the request and explain all in simple, non-technical terms. Such information may only be released to a partner with the consent of the patient. All other stake holders also need approval by the patient for the release of information. Strict rules apply to the release or transfer of sensitive data in the NHS, as indeed in any health service. Permission must be obtained from the competent authority to publish data in a research paper or book, and that too while keeping the patient's identity confidential, and unidentifiable.

3.1.2 Beneficence

This is the second pillar of medical ethics. It basically implies serving the patient for his benefit. It obviously means that the physician must endeavor to do his best to treat the patient with utmost efficiency, honesty and integrity. It is a part and parcel of the physician's psyche that he should bring comfort to his patient, by easing his pain, discomfort and agony. Obviously it goes without saying that no doctor would go another way. But the guidelines are imposed in order to define the parameters of goodness, efficiency and earnestness to relieve a patient from his sickness.

Beneficence also inherently implies that the physician would provide adequate, prompt and the finest of care to his patient. The total intention of physician should be to bring goodness through his act to the patient as indeed to the community he serves. Any form of intervention, whether diagnostic, therapeutic or investigative, must be directed towards the patient's welfare. A physician must employ and harness all his knowledge and skills to benefit the patient.

Beneficence is best translated as *Birr* in Islamic ethics. This term appears at numerous places in the noble Quran, usually combined with *Ehsaan* such as *Birr wal Ehsaan*. Both these terms carry immense depth of philosophy within themselves, which can only be discussed aptly and suitably by the learned Islamic Scholars, i.e., Ulama and Fuqaha. But as a student of medical ethics, one may benefit with the writings of a learned physician and scholar, Zaki Hasan, who passed away only a few years ago. He described that the word Birr has been used in Quran with varied meanings, depending upon the context of the message that Allah commands to convey. Birr is one of the fundamental pillars of Islamic ethics and has many implications. It does, however, transmit only one final message, i.e., truthfulness, goodness and righteousness, as indeed total surrender to Allah's commands.

Birr also means acting rightly by offering comfort to the poor, to the less fortunate, the downtrodden, the hapless, the orphans and the fallen people of the community.

Beneficence refers to a moral obligation that involved performing an act with goodness. Islam attaches singular importance to an act of kindness, and goodness condemning an evil act at all forums.

Zaki Hasan also cites an example of a debate within the Islamic thinkers on the subject of goodness amongst the two important schools of thought of the early era, namely the Ash'arites and the Mu'tazillites. The latter were the followers of Wasil Bin Ata, a disciple of Imam Jafar Al Sadiq As, who was rusticated by Imam from his academy on account of his confused thoughts. The Mu'tazalites believed that man could determine, rationally and based upon reasoning, what is good and what is evil, even prior to the revelation, in a subtle way belittling the importance of *vahi* and Quran itself. Mamun, the Abbasid caliph, promoted the Mautazallia thought, encouraging people to claim a *tahreef* in Quran. All Muslims firmly believe that not a word has been added or subtracted from this last and final testament as Allah has promised to keep it safe on the Lohe Mahfuz, away from all evils.

The Ash'arites did not agree with this philosophy and believed in strict voluntary ethics. Good is what Allah has prescribed and evil what He has prohibited. In keeping with this voluntary ethics, the Asharites were reluctant to admit that any merit was attached to the type of knowledge which is attainable through unaided reason. Allah has total control over all matters. His authority is supreme. His power is immense and sovereignty unchallenged; thus the very meaning of justice or injustice are bound up with His arbitrary decrees. Apart from those decrees good and evil have no meaning whatsoever. Therefore, one must understand that Allah is under no compulsion, as the Mu'tazzalites had argued, to take note of what moral or religious interests so to speak, but is entirely free to punish of His own accord the guilty or omit the sins if He likes.

Debates of Mu'tazalites and Asharites are well documented in Islamic literature and one may refer to the net for elaboration of the different approaches these two groups took up in their faiths.

Islam is primarily a religion of welfare and goodness. In it there is a concept of '*Shifa*,' which means recovery from an illness. It is also part of the rubric of beneficence in the context of healing or recovery from disease. Allah has mandated in Quran that He alone gives the illness as well as the relief from it: "*Wa Iza marazty fa huwa yashfeen.*" *(Alshuaara 26:80)*

The concept of God as the healer is not confined to Islam only. As Zaki Hasan argued, even the Christian concept of omnipotence of God in all matters is akin to Islamic thought, and God is indeed the source of healing and reliever of distress. In ancient Greek medical philosophy, Jupiter was the healer; thus even today most of the medical practitioners use the sign of Rx at the beginning of our prescriptions, which in fact is the sign of Jupiter. Islamic Tibb never practiced it and the *Tabibs* always began their prescriptions with '*How-al Shafi.*'

Beneficence, or Birr, was given immense importance by at the Muslim scholars and Tabibs. The patient's interest always precedes other interests. Rewards for a physician's service are perfectly lawful, and legal; but greed and exploitation is totally disallowed as that would compromise the principles of ethics beyond benefi-cence. It is of particular importance in the poor nations that Muslim countries often have. The physician must remain conscious of the fact that Allah is watching all his acts and any form of coercion, refusal to serve the poor for the lack of remuneration and discarding the elderly, the weak, the orphans and the widows because they can't pay the fees is a deplorable act. It certainly compromises the principle of Birr as it almost always goes in conjunction with *Ehsaan* which means justice with altruism, sacrifice and self denial. Islam loves Ehsaan and is a strong proponent of Birr.

One of the earliest books on medical ethics in the Islamic annals is attributed to Ishaq bin Ali al- Ruhawi. This founding father of Islamic bioethics lived in the ninth century. He gave huge importance to maintaining the highest standard of medical knowledge duly supported with skills and the dignity of physician. He believed that physicians have chosen the profession to do good and be virtuous as well as ratio-nal. He also believed that Allah chose those people in this profession because they possessed pure hearts and a sharp intellect and He chose those who love good, have compassion and mercy in their trait, added with sympathy and charity. Such people are bound to serve the patient with *Birr*.

3.1.3   Nonmalificence

It is a corollary of Beneficence. It means to treat the patient with utmost care, 'without either aggravating or indeed adding further burden of disease in the process of treat-ment.' It also implies that no false statements, overplaying or underplaying of the facts will be carried out. Serving the patient without malice, bias and prejudice is the funda-mental guiding rule, the opposite of which would be malfeasance and must not occur.

Nonmalificence may be called the opposite of beneficence, but that in terms of moral philosophy would be a negative pursuit. In a fair and just society, nonmalifi-cence should be positive and a determined act. Avoidance of harming or of aggra-vating his grief in many ways should receive priority over beneficence. This element becomes all the more relevant and important in the case of minimalism or therapeu-tic nihilism. Service may be denied, curtailed even discontinued because of the limited resources or the limitation of budgets. Such an act would amount to non-malificence. It is a common observation in the developing countries, where services are hard to come by and patients' needs demand that such services be provided; however, due to the limitations on the part of the service provider, inadvertent non-malfeasance is carried out. Now here an ethical question crops up. Who should bear the burden of nonmalificence in such a situation? Should it be the health authority, the state or the head of the state; or indeed must the burden remain with the con-science of the poor health provider?

Such a situation often arises even in the sophisticated Western hospitals. For instance, in the outpost clinics in the community, facilities are limited, so if a patient needs a certain procedure that should ordinarily be carried out on the spot, but the physician may not have the equipment or logistic help, then the service is not exactly as beneficent as it should be. The patient may either have to be referred elsewhere or given a rather subdued statement of the situation without breaking the code of loyalty to the institution, albeit withholding the correct information. It may cause some inconvenience to the patient both materially as well as psychologically. But it is not an uncommon practice. Does it amount to nonmalfeasance? That is a difficult one to reply to in the affirmative as the doctor can be absolved of his share of responsibility as he had full intention of serving the patient with beneficence but could not due to lack of facilities.

In a case of therapeutic nihilism, different faiths have different approaches. Islam is a faith of the believers. Life and death belong to Allah and we all have to return to Him in our own time, determined by no one but Allah Himself. Therefore, Islam insists that there is no scope for therapeutic nihilism and the physician must continue to do his best to save life. But Islam is also religion of compassion and mercy; therefore, once death becomes inevitable, the dying process should not be delayed. It, however, does not mean that even feeding, nutrition fluids and analgesics, or comforting drugs should be withheld. By no means should that happen. Pain relief, quenching of thirst and nutritional aids, and oxygenation must continue until the very end; otherwise, it will tantamount to deliberate malificence, even euthanasia, which is fundamentally against the Islamic teachings.

Many episodes come to mind where heroism and personal glory were the motive for unduly prolonging the dying patient's misery. Here is a particularly ghastly example that continues to haunt this author. As a first-year resident he rushed with the chief to the casualty department. The man had choked and his heart had stopped. This was well before the CPR, etc., had come into regular management strategies, but one knew how to do an external cardiac massage. It was immediately begun along with intracardiac injection of adrenaline, etc., but to no avail. Now this surgeon had just returned from the UK after gaining his fellowship and was somewhat of a hero because of his heroic pursuit in surgery; he could not take the loss of patient under his care with grace. So he did the thoracotomy on the spot and began the open heart massage, which he continued until his hands were tired. No EEG was recorded, and surely the patient had been brain-dead for a while, but the heroic pursuit to show off to the mob watching in that casualty that morning was an unethical act that cannot be forgotten. It was much more than nonmalificence, but it is best not to label it per se.

3.1.4 Justice

Justice is really the founding pillar of ethics. If one had to summarize all the cardinal rules of bioethics, one would sum it all up in one word: justice. If one observes the principle of justice in health care, one would automatically observe the code of autonomy, beneficence, nonmalfeasance and, of course, justice. Justice is the very soul of ethics. Much literature is available on the subject. Here is what I think may be relevant for our purpose due to the singular importance.

I have written a separate essay on the subject of justice, which follows hereafter.

3.2   Justice: The Most Fundamental Pillar of Ethics

Islam defines justice and a process of giving to each and every one what is his due on the basis of equity and fair play.

Two levels of justice are described in Islamic literature, namely individualistic and collective. The former has two dimensions to notice justice to one's own self and justice to a fellow human being, irrespective of faith, colour or creed. There may be two aspects of each form in its own turn, i.e., negative or affirmative. The basic intention or *Niyat* must, however, be clear, honest and dedicated to the cause of justice in either case.

At the individual level, the noble Quran defines four underlying parameters at an individualistic level:

1. To have a positive attitude and devotion to harmoniously cultivate one's own personality.
2. To be on guard against negative forces acting upon one's personality.
3. To give to others what is their due without being asked, reminded or commanded.
4. Not to defraud anyone, particularly in matters or things belonging to them.
   Quranic ethics outlines four ingredients to Collective ethics also:
1. Justice in matters of social relationship
2. Justice in matters of law
3. Financial and economic justice
4. Political justice

No doubt each one of these stems can be further elaborated, but that is beyond the scope of this book. Many scholars have written volumes on justice, which are easily accessible on the internet.

Interestingly enough, there is a metaphysical approach to justice as Zaki Hasan debated in his discourse some years ago. The ultimate justice may indeed be puritanical and metaphysical, but it can also be practical, and that is exactly what moralists and ethicists expect; i.e. justice to be done, and seen to be done. A tangible application of justice is the aim of all leaders, philosophers and scholars. But sometimes it is neither seen to be done nor indeed done per se.

Justice may not be combined or confused with goodness, as it is rather primitive concept. Goodness may be equated with being right, which means one may be faithful, with a highly moral character— but does also mean being just? Metaphysically speaking, yes, one expects a moral person to be just also, but not always. Can these attributes contribute to medical ethics? Perhaps the answer may be negative, though one expects more often than not to have at least positive influence on the final outcome, if not outright affirmative.

Another thing to remember is that social justice, which is in fact part of political philosophy, is relevant to health care and medical ethics. Islam attaches unprecedented importance to social justice, strongly condemning the unjust and promoting the cause of equity and fair distribution of resources, particularly to the needy ones, the orphans, the widows and the disabled. It is deadly against terrorism and terrorists, and strikes the tyrants with thunderbolts. The recent surge of terrorism in the name of Islam is nothing but outright exploitation, disrespect and abuse of the name. Karbala was a watershed in human history, defining the parameters of goodness versus evil. It shall remain one glaring illustration of the total violation of the fundamental pillars of humanity, namely justice of Islam. They are not even human beings, let alone do they have any form of faith let alone Islam. They are animals in the garb of human beings, who have hijacked Islam through the petrodollars of the Arab world. Such a disgrace to the land where Islam came into being is a blatant illustration of how unkind the so-called followers of Mohammad have been to his homeland. But what is that in comparison to the annihilation of the family and the household of the Prophet, which was meted out soon after the departure of the Prophet form this world.

Modern bioethics is partly a fall out of the Second World War. The Nazi-era experiments on the inmates of the various camps, followed by the Nuremberg Trials, must have acted like a watershed of some sort to precipitate the matters, as they stand today.

It was not the Nuremberg Code only but the Civil Rights Movement of 1960s also, compounded with rapid expansion in the knowledge as well as the skills of scientists, which nurtured the concept of re-evaluation of the paternalistic approach of the physician that has been part and parcel of the profession from innumerable centuries in the past.

It appears that we may be witnessing only be the tip of the iceberg, as with the new discoveries in medical technology, no one can predict the final frontiers that science may reach in the future. Perhaps, after all, the medical profession may be obliged to practice what it has preached from the days of Galen and Hypocrites.

Human experimentation did not end with the infamous Dr. Mengele and his unholy team of scientists. The process continues to remain in practice more than half a century after the war.

The events of September 11 shall go down in the annals of human history as one of the tragedies of modern times. The aftermath of that episode continues to be felt across the globe in seismic proportions. What followed the September 11 phenomenon was the Anthrax scare. The American media as well as competent authorities instantly raised their finger toward the already established culprit. Immediately all eyes were looking at Iraq.

However, soon it was revealed that the lonely scientist sitting in a remote lab within the United States might have been the initiator of the whole scary business. In

Now we turn our attention to the matter at hand. Human experimentation is perpetually being carried out without an informed consent in many parts of the world. In fact, the concept of informed consent is quite new. It was only in 1972 that it began to receive any form of detailed examination. There are many physicians who are quite oblivious of the subject even today. Hence, knowingly or otherwise, many of us may get involved in one form or another of human experimentation in the deprived and the underprivileged countries, mainly in Asia and Africa.

We shall quote some of the salient examples of such unholy practices in both the developed and the developing countries.

The more recent episodes of American involvement in human experimentation is the revelation of such clinical trials without informed consent and under risky conditions, as the CIA's MKULTRA program and the American Army's LSD experimentation, disclosed after the Vietnam War and the Watergate scandals, respectively. And only a year ago occurred the surfacing of human radiation experiments carried out upon men and women in uniform on the British and the American forces in the Gulf War. The final implications are still under scrutiny.

Earlier on, there were several major milestone stories that shook the world and are worth recapitulating.

1. The United States Public Health Service funded the now so-infamous Tuskegee study in the state of Alabama carried on for three decades, from 1932 to 1972, to investigate the effects of untreated syphilis on 400 black males. The null hypothesis under test was that it was "the treatment of syphilis with a new drug more effective than no treatment at all."

   The obvious harm in the investigation was to withhold treatment with penicillin without an informed consent of the subjects under trial. In 1997, President Clinton had to formally apologize for the abuse incurred during that generation's long study [1].

2. The second example is that of Baby Fae, who could well be called a pathetic victim of the so- called research.

   The details of this experiment are well documented in the literature, including the rapid demise of Baby Fae and the pathos it created for the single parent, i.e., the mother.

3. The AZT experiment: The results of the AIDS Clinical Trials Group (ACTG) study 076 revealed that in France and the United States, vertical transmission of HIV from pregnant mothers to their neonates could be reduced by two-thirds (25.3–8.3 %) with an intense regimen of AZT therapy. The trial was then halted. Six months later the WHO held a meeting of intra-national researchers and health officials in Geneva, following which the CDC and the NIH of the USA embarked upon collaborative studies on pregnant HIV-infected mothers in Africa and Thailand. The experiment involved a placebo arm so that the control group would receive an inert substance, despite the proven status of HIV [2].

   The question of whether this trial could be conducted anywhere in a developed country in the Western world needs no answer. In fact, only today the BBC radio was interviewing the South African health authority, challenging their decision to disallow any such investigation in their patients.

4. The Willowbak State School on Staten Island (1956), which involved deliberate infection of mentally disabled children with hepatitis.

5. The cancer programme at the Jewish Chronic Disease Hospital (1963), involving deliberate injection of liver-cancer cells into patients, obviously without an informed consent [3].

6. The testing of unapproved drugs on military personnel during the Gulf War in 1991.

7. The story of enriched uranium used in some military arsenal employed by the military personnel during the Gulf War. And the list goes on . . .

   These examples are just a few to highlight the point that the developed countries, in their best interest, can play with the lives of their own 'underprivileged' people without being squeamish about it, so to what extent can they go in the developing countries remains open to one's imagination!

3.2.1 Is Human Experimentation Ever Justified?

Well, in order to investigate that, we need to momentarily refer to the philosophical theories relevant to morality and ethics.

Two theories have mainly dominated the scene from about the mid-eighteenth centuries in Western philosophical literature. However, we shall also look at the more ancient and divine theories ordained by Islam fourteen centuries ago.

Kant's Philosophy of Deontology

It is also called the Obligation Based Theory. Immanuel Kant was a German philosopher who lived in a small town in Germany between 1724 and 1804. He is said to have lived a life of utter simplicity, mundane contentment and self-sufficiency. However his works have left an indelible impression on Western thought.

The theory of Deontology puts forward the concept that "some features of actions, other than or in addition to consequences, make actions right or wrong." It therefore implies that an action does not depend solely upon its consequences, since there may be certain features of the act itself, which determines whether it is 'right' or 'wrong.'

Kant firmly believed in human autonomy and the dignity of mankind. He expected a 'good' man to do nothing but good to his fellow beings. He believed in the 'rationality' and 'reason' that a man is endowed with as opposed to animals, who are simply governed by 'desire.'

Kant agreed that morality is grounded in pure 'reason' (*Idrak*), and not in tradition, intuition, conscience, emotion, attitude or feelings such as pity, mercy, empathy, sympathy, jealousy, animosity, hatred, etc.

Kant also believed in the ultimate supremacy of mankind over all other creatures. He believed that a man must not act in response to desire or indeed as an act of 'obligation.' He must act on the basis of his 'will to do' and because he considers an act 'necessary to do,' not because it is his 'duty' alone and so he must comply.

His fundamental thought was that 'one must act to treat every person as an end and not as a means' only.

Bentam and Millers' Theory of Utilitarianism

It has been discussed elsewhere, but suffice it to say that this theory informs us that the consequences of an action are more important than the act itself. Therefore, in certain situations even a less than good act applied on an individual in order to save larger population from an evil act may be justified. Animal experimentation is justified on the basis of this theorem, as by sacrificing an animal one is helping save a human life. It obviously is debated by pro-animal life activists who have raised many an objection to animal experimentation in the UK. Basically the theory says that the end can justify the means.

Some workers have employed the Utilitarian thought to justify human experimentation. But they tend to forget that so did the Nazi warlords when they justified their human experiments on the basis that for the benefit of larger group of people, a minority may be exposed to harm. However, this theory is outright rejected because the essential component of any human involvement is the concept of informed consent.

Of course, the German Nazis happily employed the Bentham and Miller's philosophy on the basis of its utility for the benefit of mankind. They used the human subjects as the 'means' to achieve an 'end'—in their case, what may be called the 'double effect'—getting rid of the unwanted human population while serving mankind through 'the so-called research.'

Imam Khomeini was a great reformer and ethicist. His famous book on ethics made the father of Reza Shah go rabid, and caused many untold hardships for the Imam. But Imam Khomeini was not alone in writing on ethical issues. His followers and many other Muslim philosophers have also contributed to ethics and bioethics over the past several decades. Many other ethicists of the Muslim Ummah have written extensively on such subjects as are contemporary in nature. They have also introduced new terminology and concepts to either apply them in practice to solve a dilemma or for other authorities to ponder over. *Maslaha* or the 'doctrine of public interest' is one such term often used to look at the Utilitarian concept of Western philosophy, in such matters where larger public interest may justify a given action on a person.

Beneficence is best translated as *Birr*, but many have also explained it on the doctrine of 'la *darar wa la derar,*' basically meaning, 'do no harm.' That, in fact, is the soul of beneficence.

Another word coined by Muslim ethicists is *Darura,* which means 'necessity,' employed in certain situations where a necessity may have arisen to perform an act obviously with the motive of doing good to the patient. Similarly a term called *La Haraj* means that 'no hardship' be inflicted in the matter of patient care.

Frankly, it is rather unfair of non-Arab scientists, thinkers and ethicists to comply with the dictates of Arab ethicists as many terms are equally, and adequately available in other languages too. It is not the word of God nor a *hadith* from the Prophet that can never be altered, but the force of circumstance is such that despite a majority of Muslims in the world being non-Arab, others are obliged to use the Arabic terms. It does create confusion and some misunderstanding as the meaning of the same word may be interpreted by a non-Arab Muslim quite differently than an Arab scientist. In fact even the *Ulema* of different ethnic backgrounds have noted such interpretational differences in their respective fields. It is therefore necessary that uniform guidelines and nomenclature must be identified by the Muslim *ummah,* which should be applicable across the board.

Justice is either retributive or distributive. The philosophy of reward and punishment is as old as time itself. Religions have always warned that there is an inevitable return to come forth in the life hereafter, for our worldly deeds. Hence, all the divine books have provided their followers with the road maps to follow the right path avoiding pitfalls and destruction. Islam has given clear mandates that there would certainly be rewards for good deeds, but one cannot escape the ultimate punishment, a lasting place in the abyss of hell, for the evil deeds.

Justice is one of the four pillars of Medical Ethics described by Beaucahmp and Childress [3].

Justice determines that the equitable and fair distribution of services may be dispensed to all without any bias of race religion, colour or creed. The theory of Utilitarianism has many faults as has indeed the Kantian philosophy of Deontololgy.

One of the criticisms leveled against Utilitarianism is that it does not take into account 'Justice' and definitely does not secure an equal distribution of 'happiness,' an entity considered to be a definition of ethics which need not, therefore, deal with the individual according to his or her desire or indeed merit.

Thus, a court believing in Utilitarianism may indeed be quite justified in condemning an innocent man to death if it was considered that a greater good would result from such an act, such as restoring law and order, etc. Obviously this concept raises many questions. Justice could only be delivered if all matters are weighed upon the scale of unbiased decision.

Regrettably, the Utilitarian theory is wrongly interpreted in many countries. Many times scientists employ this theory and thus carry out human experimentation firmly believing that a greater good will come out at the cost of some harm to a few.

Unfortunately, Utilitarianism and analytical ways are often said to fail to take into account the problem of Justice, because they focus on the net balance of 'benefits' over 'costs' without considering the distribution of those benefits and costs [4].

Some scientists seem to believe that since the developing nations are too poor to afford any form of treatment at all, an experimental or a therapeutic trial would at least grant them some access to treatment!

They fail to appreciate is that distributive justice needs to be put to the test on the balance of 'risks' and 'benefits.' As we shall see during the theoretical elaboration of the theories of Distributive justice, the 'Risks' would be entirely borne out by the population under trial, and the benefits—most if not all to the 'advantage' of the nations who bear the 'cost' of these trials.

Beauchamp and Childress, while discussing the possible patterns of distribution of risks and benefits, outline as

1. The risks and benefits falling on the same party.
2. One party bears the risks, while another party gains the benefits.
3. Both parties may bear the risks, while another party gains the benefits.
4. Both parties gain the benefit while one party bears the costs.

In most clinical trials in the developing countries, the second pattern appears to be more often the case. However, in a few circumstances the fourth pattern may be seen at least at the beginning of the trials, as was the case in the AZT experiments.

The term 'distributive justice' refers to fair, equitable and appropriate distribution in a society determined by justified norms and structure in terms of social cooperation. Its scope includes policies that allot diverse benefits and burdens, such as 'poverty,' 'resource taxation,' 'privileges' and 'opportunities,' etc. .

The term distributive justice is sometimes also used broadly to refer to the distribution of all rights and responsibilities in a society, including such rights as the civil and political rights, right to vote freely, freedom of speech, etc..

The problems of distributive justice arise under the scarcity of resources and a climate of competition.

The following principles have been proposed as valid materials for distributive justice:

1. To each person an equal share.
2. To each person according to the need.
3. To each person according to the efforts.
4. To each person according to the contribution.
5. To each person according to the merit.
6. To each person according to free market.
7. To each person according to exchanges.

Most experts such as Beauchamp and Childress have described the following influential theories in respect to Justice.

1. Utilitarian theory: It emphasizes a mixture of criteria, for the purpose of maximizing public utility.
2. Libertarian theory: It emphasizes the rights to social and economic liberty (invoking fair procedures, rather than substantive outcomes).
3. Communitarian theory: It stresses the principles and practices of justice that evolve through traditions in a community.
4. Egalitarian theory: It emphasizes equal access to the goods in life that every rational person values, often invoking material criteria of need and quality.

Islam attaches singular importance to justice. In fact, one of the most sublime attributes of Allah is 'Al Adil,' meaning thereby 'The Just.'

Justice is one of the attributes of perfection, and God is indeed the ultimate and the supreme power that controls all matters in life as indeed in death. He commands people to dispense justice and strictly prohibits injustice [5].

In the holy Koran in Sura Nisa [6], Allah commands, "Allah does not do an atom of injustice (4:40), then He commands [7]" (18:49) "Not one will your Lord treat with injustice".

In another Sura [8] Allah says, "Allah does not will injustice to His servants" (40:31).

In Sura Nisa [9] He once again commands, "Whatever befalls you is from your own doings" (4:79).

And in Sura Sajada [10] Allah gives a clear mandate that "[It is] He who made all things good that He has created" (32:7).

There are numerous definitions of justice. One such definition says it is "giving one what is his due on the basis of equity" and justice. It, however raises the basic question as to whether equity and justice are interchangeable terms or not. Personally, I think they are two different entities. Equity implies equitable and perhaps equal distribution amongst the recipients. Justice, on the other hand, may not grant equal share to all, as some cannot deserve any reward whatsoever, and others may have huge portions. Allah is just. If He was equitable and not just, all His creatures would have received equal benefits, but that is not true. One gets reward according to and relevant to his deeds, as does he receive punishment for his sins matched to the crime committed. If the reward was more than duly deserved or punishment excessive than what is one's due, it may not be a just act. Allah grants favors and punishment with justice and not equity per se.

The holy Prophet is quoted to have said that "one act of justice is superior to a thousand *Ibadats*" and Imam Ali As, the greatest jurist of all times, defined justice as "orderly placement of all matters," the opposite of which, of course, is chaos and is called *Zulm*. Imam Ali also gave us the founding pillars of justice in one of his sermons as profoundness of knowledge, depth of understanding, foresight and prudence. His justice is well documented in the annals of Islamic history. It is perhaps rightly said that one of the reasons of his martyrdom was that the rich and the powerful of the Muslim *Ummah* feared his retribution and could not tolerate his justice.

Justice is a pivot of any welfare state and if the state cannot serve its people with justice, it has no right to rule. Imam Ali also said that an infidel's rule may last for sometime but not that of a tyrant. Haven't we seen exactly that in the episodes since 9/11, that the tyrants in Iraq, Libya, Egypt and elsewhere could not find a safe abode even in the drain pipes or the rabbit warrens. And yet mankind learns no lessons from history. Tyrants continue to rule in many parts of the Muslim world, depriving their subjects of the fundamental human rights, let alone granting them justice, all in the name of serving their people, even Islam!

Most societies in the world call themselves just, even if they perform questionable acts. But none boasts about being unjust, as they all know the true significance of justice and its impact on the human mind. In the history of mankind, however, there has been time when a form of government existed, which was proud of its tyranny and lack of justice. These rulers were proud of themselves and of being addressed as the Tyrants. But if you ask even a most well-read student of history, he would not be able to recall even one name of the tyrant member of that dynasty. In contrast, even a lay man can tell you that there was a ruler in ancient times called Nusherwan, whose title was *Adil*, because he was just. Such is the contrast between justice and the opposite of it, i.e., tyranny.

Insecurity breeds greed and fear begets cruelty. Justice is the only barrier between the good and the evil. While distributive justice is a popular theme, one may not forget its exact opposite.

Retributive justice is the other aspect of Justice. It primarily means that a crime must be punished in equal measure. It is important that the retribution must match the crime committed; neither should it be excessive nor less than it ought to be. The *lex talionis,* or 'law of retribution,' is practiced in many cultures. It basically means that 'the punishment must equal the crime.' In fact, Quran has given a fine illustration of retributive justice when it commands the *momineen* to take an eye for an eye and a limb for a limb. The ultimate crime is a murder, and its punishment in the eyes of Islam is a death penalty; however, there is clear concept of *Qisas* and *diayt* in Islam. Compassion always takes precedence over punishment. The finest example of this concept was witnessed at the time of *Fateh* Mecca, when the enemies of Islam, including the murderers of Syedul Shuhada Amir Hamza, were brought as prisoners of war in front of Prophet, expecting nothing short of total annihilation for all the evil acts they had committed against Islam, but Prophet forgave all his enemies including the murderers of Badar and Uhd, setting an example for us all to forgive rather than avenge.

In fact, many ethicists in the West who are dead-set against a death penalty claim that the criminal may yet reform if left to suffer all his life in a goal rather than be killed as the practice in many countries. Well, the pro-death penalty group claims that the goals have failed to reform the criminals; in fact, worse may happen to them when living with hard-core criminals. But that is another debate.

It sounds all very good to talk about justice and its importance, but do we see it being practiced today as it is preached; or indeed, are they all hollow slogans that sound endearing, charming and mob pleasing? Well, the truth of the matter is that justice has always been crushed under the feet of the rulers throughout the history of mankind. And yet its significance has all but diminished. There have always been a few daring people who have stood the massive onslaught of tyranny, facing destruction and death and opposed injustice at the cost of their home and hearth. Such are the folks who have become the legends and true heroes of folk tales.

In the twenty-first century, one is appalled to see the examples of injustice being constantly handed out to the poor and the downtrodden. The tales of Charles Dickens era are old, but are they? You may yet find an Oliver Twist and Fagan in any society, anywhere in the world. Nothing has changed, but now that the media is so powerful, such stories see the light of the day much faster than in the previous centuries.

All philosophers and all religions have taught the same thing; i.e., support and practice justice and condemn the lack of it, even if your own life is at stake.

In the field of medical ethics, what we see today is very disturbing. The rich and the famous have the benefits of access to the finest of health care, but the poor and the needy die on the roadside. The richest nations are the worst in that way. The United Kingdom is one country where the general rules of equivocal distribution are practiced throughout the health services. But the NHS is under constant pressure by the mangers and the financial wizards to cut down the costs, Therefore, we have seen so many hospitals either close down or at least have their services curtailed to meet the financial targets set up by the government.

The year 2006 was badly hit by the financial constraints, when many hospitals trusts went into the red, and had to shed their staff and cut down many a service. The gloom and the depression that hit the nursing, medical and paramedical staff last year across Britain was very hurtful indeed. In many hospitals petitions were signed in favor of old staff about to retire to at least be allowed to earn their pension, requesting from many years of service in the trusts to allow only a few months to retire gracefully rather than face the unpleasantness of the sudden removal from many years of honest work. It was all political. The labor Government of Tony Blair believed in meeting certain targets, both fiscal as well as number of patients, and cutting down on the waiting period. It compromised the services to a considerable extent. They were hues and cries from the opposition benches as indeed from the professional pundits. But the stubbornness of the politicians saw to it the targets were met, even if people faced misery. It was far from the fundamental rule of justice to cut down on the treatment as well as the staff number providing services to the elderly population in the nursing homes as indeed in main hospitals. The process hasn't stopped so far, i.e., until the summer of 2007, though the trusts are now mostly out of the woods.

Rationing of health services has always been in practice, as no country rich or poor can provide all services to all of its citizens. But the form of rationing seen last year in Britain, besides many other reasons, may be the reason for the departure of Tony Blair. There were all kinds of complaints including the lack of supply of medicines. It may surprise many readers in the poorer countries, but even in the mighty Britain, life-saving drugs are sometimes not given to the patients on the NHS. The same kinds of constraints are applicable to newer drugs for Alzheimer's and similar crippling states.

One elderly patient this morning told me that he was so fed up waiting for a hearing aid through the NHS that he had to buy one privately for £550, which as he said could have bought him a new bathroom. And there are many similar stories for patients being either too tired of waiting to receive the services or simply deprived of drugs or treatment.

Since the concept of NHS was, and remains to date, that each citizen of the UK is entitled to free and fine medical care, the concept of rationing is somewhat a later day entry. The elderly patients inform us that before the advent of the NHS, each one had to buy the health care himself, and since the economic conditions in many parts of England and Wales were very bad, many could not afford any form of treatment and simply succumbed to the illness. Even today one can see the scars of the olden times in places like Rhonda Valley in South Wales, a place where the coal mines were killed by the thousands. TB was the major killer and little or nothing could be done to help those unfortunate patients. Old sanatoria have now been replaced by scenic parks and touristy places as they were often located away from the main cities, mostly in the country side.

The cities of Lancashire, like Bolton, Blackburn and Preston, still show the telltale signs of textile workers who developed COPD and were treated in small hospitals across the width and breath of these lands of the workers. But all that has changed gradually, and the modern cities like Manchester and Liverpool are totally different from what many of the older generation saw in late 1950s and 1960s.

The current situation in the UK is that since it is no longer feasible to provide all services in all hospitals, which were called district hospitals or infirmaries, it is best to consolidate most services under one roof on a regional basis and provide day-to-day service through the community heath services, focused around the general practitioners and the Primary Health Trusts. The new mega- hospitals being built nowadays will be the state-of-the-art centre for extended medical care. They are expensive, and controversial in the eyes of some cynics, but the experts inform us that since the technology has evolved so much, you simply can't have the pre-war infirmaries coping with the latest developments.

Another factor that has expanded in recent times is the concept of day surgery. Most procedures are now performed on the basis of day surgery, which is extremely cost-effective. No wards means no 24-h monitoring, no nursing, no botheration of inpatient management, etc. And then there is the category of a 23-h stay, i.e., less than 24 h, where one may require observation and overnight care. Both these techniques are extremely cost-effective, fine examples to follow in the developing economies. But it is not a new concept at all. I remember some of my teachers practicing day surgery on their patients. They were mostly cases like Tonsillectomies and Adenoidectomies, etc. The principle in their minds was obviously the same as today's NHS mangers: i.e., cost-effectiveness and less hassle.

Services within limited resources are the standard management slogan. To this, a fresh modification has been added in the past few years and that is 'within limited resources.' The whole concept revolves around the philosophy of providing at least the basic services to all and, where possible, most services to whoever needs them; but when it comes to the highly specialized and therefore expensive services, some form of rationing would be inevitable. And that is the crux of the matter, as to the criteria upon which one would choose one patient against the other. Lately there has been considerable debate in the UK on the introduction of choose and book practice and also the choice given to the patients for selecting their doctor and the hospital. It has raised many questions, and some trusts with rather questionable record are raising objections. The fear is that, given the choice, the patients may not come to them and go elsewhere. That would obviously mean a fall in the revenues and financial problems.

No progress in medicine is possible without research. Unfortunately most of it is carried out in the developed nations as they have the financial resources. But since money itself cannot produce results, the scientists also need manpower to experiment on. And that is where the developing nations come in, regrettably though not often as joint partners, but as the victims of the circumstances.

Animal experimentation is nearly always carried out in the rich countries. In fact, this issue by itself has created many problems for the scientists in recent times. The suppliers of animals for experimental works in Britain have been the targets of pro-life protestors and animal lovers. Last year several homes and farms where guinea pigs or hamsters, etc., were raised were attacked by the protestors, some in close proximity of Oxford and Cambridge. The suppliers were harassed to the extent that some had to abandon their profession altogether.

The same formula, though, does not always apply to the human beings who are often employed as guinea pigs. It is happening all over the world. China and Latin America top the list, but India, Pakistan, Sri Lanka, Bangladesh, Nepal and many countries in Africa are the perpetual victims of the experimental use of human beings. The commonest modus operandi for a renowned Western University is to develop a so-called collaborative programme. Not always, but sometimes the chain of command stretches from the top man at the health and medicine affairs down the line including the academia. Perhaps some aspects are absolutely ethical, but some are obviously questionable, such as the basic principle of informed consent, or confidentiality.

The most troublesome aspect of such trials is that once the phase-four trials are satisfactorily completed, the drugs are withdrawn from the very people on whom they were tried. That is not only beyond the realms of justice but indeed outside the domain of humanity. It has to be condemned and must not be allowed by whomever is involved in such heinous work. Distributive justice demands that once the trials are over, not only must these medicines be made freely available to the local population, but charity demands that they should be given free of cost.

Greeks liberated the human mind, but promoted, endorsed, even encouraged slavery. Islam taught, preached and practiced just the opposite. It not only liberated the human mind but also mankind itself from the centuries-old tradition of slavery and human bondage.

Greek greats formally taught ethics, and defined the parameters as human dignity, individuals' rights of freedom of thought and speech, the virtues such as truth, and justice, etc. In fact, Socrates sacrificed his life to uphold these very principles and Aristotle gave his virtue theory, which has stood the test of time. Justice played a singularly important role in the Greek philosophy.

But no example in human history can match the sacrifice of Imam Hussain AS, who in Karbala, sacrificed everything, to uphold justice, truth, honesty, integrity, faith, and save Islam from total annihilation at the hands of Banu Umayyah. His example is outstanding and gains ever-growing shine and glory as the truth behind his massacre is unfolded each passing day. The obvious question that comes to mind is that who, why, and with what authority this great human being was murdered and his family treated with utmost indignity. The world shall one day realize that Hussain was not just a name, he was the epitome of mankind, a symbol of total goodness who rose up against total evilness a long time ago, but whose memory and revival of his sacrifice is gaining strength each passing day. Chosen people of Allah are never annihilated by tyrants, but only temporarily tormented or eliminated. But history never forgets them, and they shine through the pages like a glow of bright and heavenly light in total darkness.

Islam has served the cause of scholarship through the ages. When Europe slept in total darkness, Islam was serving humanity through its illustrious sons, who perfected the arts and sciences in a thousand fields. Between 610 AD, that is the dawn of Islam, and 1492 AD, when Muslims lost their glory through petty jealousies and infighting, Muslims covered almost 800 years of human history. After the fall of Grenada, the Ottoman Turks took over the reins and spread the faith and its scholarship into the Indian subcontinent. India, in its own turn, produced many scholars, judges, philosophers, artists and teachers par excellence. It was only after the fall of the Indian empire at the hands of British Imperialists that Muslim India succumbed, though elsewhere they continue to produce the likes of Rumi, Hafiz, Saadi, Mir, Ghalib, and Iqbal.

The new world order has pushed the Muslim *ummah* to its farthest edge, but scholarship continues to prosper. Islam, which began in Mecca, did not remain confined to Mecca, Kufa Baghdad or Cairo. It has now engulfed the whole world and there are more non-Arab Muslims in the world than Arabs. That means all races and all continents are following the teachings of Mohammad SA, whose message was the divine command in the form of *Iqra*. The message continues to serve the world. Prophet himself was the ultimate teacher of ethics, i.e., *Moallame Akhlaq*. His teachings were transmitted to us through his family, who continue to guide us in our daily lives.

His grandson, the fourth Imam Syed e Sajjad As, has left us a collection of his supplications. They have been compiled into a book called the *Saheefa e Kamila*. It has one particular supplication fully dedicated to the normative principles of ethics. It is called the *Daue Makram* al *Akhlaq*. As a physician, I have drawn a vast number of lessons from it and have documented it in an essay in this book.

Medical ethics is growing rapidly, and the physicians are not quite prepared to solve many a dilemma, due to lack of full knowledge of principles of ethics and their application. It is necessary that the medical profession be thoroughly knowledgeable about the subtleties of medical ethics, particularly the principle of justice, in dispensing medical care.

References

1. Plamer M (1999) Moral problems in medicine. The Butterworth Press, Cambridge, pp 82–83
2. Planner M (1999) Moral problems in medicine.The Butterworth Press, Cambridge, p 88
3. Moazzam F (2000) Human subjects research ethics and the developing world. J Pak Med Assoc 50(11):388–393
4. Beauchamp TL, Childress JF (1994) Principles of bioethics, 4th edn. Oxford University Press, New York, pp 315–330
5. Sayyid Mohammad Hussain Tabatabai (1979) Basic teachings of Islam. International Publishing Co, Tehran, p 55
6. Quran, Sura Nisa 4:40
7. Quran, Sura Al-Kahaf 8:49
8. Quran, Sura Al-Mu-min 40:31
9. Quran, Sura Nisa 4:79
10. Quran, Sura Sajda 32:7

Social Anthropology

4

4.1 Impact of Culture on Ethics: and Moral Relativism

Every country, every nation, every race has its own ethos, which forms the basis of the ethics practiced in its day-to-day life. In order to study the nature of Indo-Muslim bioethics, we must first study its culture, traditions, norms, values, indeed the very code of conduct that is employed in day-to-day matters. In other words, let us carry out a study of Indo-Muslim social anthropology.

India has existed from ancient times. It has a long history of its traditions, culture and civilisation. Muslim influence in India dates back to early excursions in the times of Caliph Umar. But the main invasion of India and particularly Sind happened at the orders of the tyrant Hajjaj bin Yousef, when he despatched Mohammed Bin Qasim to take revenge from the Hindu rajas for an alleged maltreatment of Muslim women. Only a couple of decades earlier the Hajaj's masters had treated the ladies of the Prophet's household with utmost disgrace. Of course that was not to be mentioned in those days of Banu Umayah, who demolished the family of the Prophet at Karbala and Sham in 61H. Syed, an eminent Sindhi leader, wrote that Hajjaj and his masters actually despatched their forces to chase the ladies and a few men of the friends of Ahlalbeit, who were sent by Imam Zain al Abedin to Sindh because the Imam had some family connections with Sindh through his wife by the name of Jawareiah. This lady was probably from Thatta and was presented to the Imam by Amir Mukhtar Saqafi.

The *Rjahas* of Sindh provided shelter to these immigrants, just as Negus (Najashi) of Ethiopia had done for Jaffar ibne Abu Talib and the earliest immigrants of Islam. Therefore, a historical misreporting that Mohammad bin Qasim was a part of tyranny and Raja Dahir a saviour of the immigrants supporting the Sadat must be rectified. Syed questioned, in the light of these facts, who should therefore be called a hero?

Let us look at the later-day developments brought in by the early Muslim invasion of India. As we know, Pakistan came into being on the basis of 'two-nation theory'. In fact, some diehard supporters genuinely believe that this theory traces its origin back to the day when the hooves of the first Arab horse shook the mortal lands of Sindh.

S.H. Zaidi, *Ethics in Medicine*,
DOI 10.1007/978-3-319-01044-1_4, © Springer International Publishing Switzerland 2014

Pakistan is a land of the pure – or at least so the word *Pak* means. Technically speaking, it is only 64 years old, but historiographers inform us that its history is several thousand years old. Churchill once said that you should look into the past to see into the future; hence, one can't simply look at the 60-odd years, which are only a tiny moment of existence in the life of a nation that has as ancient a civilisation within its boundaries as the mighty civilisations of ancient Egypt or Assyrians or the Incas.

It is a melting pot for many an ancient culture and civilisation, with strong links across the valleys and mountains of the Himalayan range, indeed even well beyond the mountains and rivers of Asia Minor, and vast lands of the Arabian Peninsula, India, Iran and elsewhere. Its roots may well stretch from the Caucasus Mountains, the legendary *Kohe-Qaaf* of the famous Amir Hamza, *Tilsme hoshruba* and *Hatam* Tai tales to the wilderness of the Kohi-Nida.

Pakistan has a multitude of races that constitute its proud people. There are the ancient Indo-Aryans, Scytho-Dravidians, the Sakhas, the Turko-Iranians, the descendants of Arabs and Afghans called the Indo-Arabs, the ancient Caucasusians, Mongoloids like the Hazaras, and the formidable Negroid races called the Shidis, a corruption of the word *Seidi*. And then there are those natives who converted to Islam in the days of Muslim invasions, albeit mostly through the preaching of saints and *derveshes*. This mostly happened in Sindh and Gujrat, but to a certain extent in Punjab also. The Khojas, who were actually Khwajas, somehow traced their connection with Khwaja Abu Talib and the Aga Khanis, who converted to Islam, and mainly to the Shia faith, in 1870s. The telltale signs of their heritage are visible in surnames like Dharamsee, Laljee, or Najianis.

They all have powerful cultural values, which make an interesting and quite daunting study of the anthropology of Pakistani races and the impact of their cultures on bioethics. Pakistan is indeed a cauldron, a melting pot of many an ancient civilization – an ideal case for the study of the impact of cultural integration (or the lack of it!).

National Geographic is an illustrious magazine. In its August 2007 issue, the lead article was written about Pakistan. It is an absolutely must-read for any Pakistani. One may disagree with some of the writer's observations, but one is obliged to give kudos to the opening paragraph, which nearly sums up the cultural anthropology of Pakistan.

The author defined the Marghala pass, barely 16 miles north of Islamabad, as the seismic centre point for the so-called clash of two civilizations. North of this pass is the mountainous terrain of the Hindu Kash, the Karakorums and the Himalayas that have kept the hardy Pahstoons and their puritanical form of Islam from the low lands of the Indus valleys, incorporating thousands of years of ancient civilizations, stretching from the Taxilla and Harrapa to Moenjo Daro. The conflict has raged for centuries between the rocky inhabitants, where the lands are uncultivable and the weather hostile, and the soft and milder nations living in the fertile lands of the Punjab and Sindh; both providing shelter to the hard-working farmers and peasants, feeding the armies of the invaders from the north, as well as serving as a food basket for the Indo-Pak territories over millennia.

The cultural impact of a nation or a country on moral philosophy and the regional practices of norms remains a strong point. Each civilization may have many colours of culture within itself, akin to a kaleidoscope. Islamic civilization is an excellent example here. Islam has universal teachings for all its followers across the globe; however, its practices have been affected by the local culture of the country where it is practiced. For instance, an Arab Abaya in the cold climate of the Scandinavia is simply impractical, as is wearing a Parka in the midst of the Sahara, but the format of salat is universally the same as practiced in the Holy Kaaba. Similarly, the customs of each country and, even within a country, each tribe may somehow affect the norms. That is why in order to understand the bioethics of Pakistan, we must look into the chapters of history and geography as indeed social sciences to understand the norms of morality and ethics in Pakistan.

What has been described by Childress and Beauchamp in their book called *The Fundamental Principles of Bioethics* is absolutely wonderful but cannot be unequivocally applied to all nations. However, that is another debate as to whether one believes in universalism or relativism.

No doubt, certain principles of morality are universal and cannot be altered or rejected to be applicable across the cultural divide, such as truth, honesty, bravery, generosity, etc., but within these values there could be many colourful shades that reflect the cultural impact.

Moral relativism is often accepted by many ethicists as the only plausible alternative to the concept of universalism. It is defined as a philosophy in which norms and values vary in many ways and depend upon the cultural variations. In other words each culture and of course each civilisation has its own moral values, which are suitable to them but may not be acceptable to others. Many philosophers strongly advise that cultural diversity and moral norms are quite compatible and indeed more practical than the concept of universalism. The subject deserves detailed discussion.

On August 14, 2007, Pakistan reached the age of superannuation. It is now senior enough to merit serious attention of a world-wide audience to what it says and does. The world has never paid so much attention to Pakistan in its entire rather adventurous history as it has done since 9/11. More so since May 1, 2011. Pakistan is indeed going through very challenging times, but such has been the history of Pakistan that, despite all odds, it has survived and shall live Insha'Allah forever.

Some pundits say – quite rightly too – that Pakistan was hastily carved out, at the command of Lord Louis of Burma, by a man called Cyril Radcliff (who had never ventured farther than Paris in his entire life and knew nothing about human trauma) on a piece of paper in his closed quarters. Everyone hoped that it would succumb in a short while. It has, however, outlived the mighty Soviet empire. And that is no mean feat for a nation to achieve that has been plagued forever with external as well as an internal Nawab Jaffars and their clones.

Sirajudullah and Teepu Sultan suffered humiliation and defeat, not because of any lack of courage or bravery on their part but the treacheries of the toddies bought out by the cohorts of Clive of Drayton Manor, a tiny village in Shropshire, in previous centuries. Till today, even in his own backyard, not many people admire Clive, and his statue is erected in London, facing away from the Parliament houses as if to display his disgrace at the hands of the British Parliament, despite his so-called glories in India. Evil does not ever prosper though temporarily it may reap rich rewards of its acts, but only temporarily.

On a visit in February of 2013 to Edinburgh, I saw the famous painting in the National Art Gallery, on the Princess Street, in which a Scot by the name of Brynt stood near the body of a fallen warrior, as the painting records, with the caption that the "the body of Sultan Teepu Saib was discovered at the siege of Surringapatum on 4 May,1799." That picture is so gripping that one just stands there looking at it, while the history of past two centuries of invasion of India by the British runs through the mind like a film. If only Teepu Sultan had not been betrayed by his subordinates, and the Nizams of Deccan, the Indian history may have been different. Teepu was a Shia. It is also obvious from an arm band, an amulet, called 'Imam Zamin,' wrapped around his upper arm, so clearly visible in this painting.

Watching that painting in the Edinburgh museum, almost standing in a trance, an email ran through my mind. It was sent by an unknown writer, following the merciless killings of more than a hundred Hazra Shias in Quetta in early 2013, followed by bombing of a Shia neighbourhood called Abbas town in Karachi. It said that all these Shias were killed as heretics, so if I have to prove to the present day Pakistanis that I am a Muslim, the writer wrote, I must call all these people heretics. And the list that the writer gave included the names of Mir, Ghalib, Anis , Dabir, Sauda, Syed Ahmad, Hali, Azad, Mohsin ul milk, Mohammad Ali Jinnah, Agha Khan, Raja Sahib Mahmmod abad, Iqbal, Josh and saints like Shah Lateef, Shabaz Qalandar, Sachal Sarmast, Madhu lal Hussain, Abdullah al Ashtar, and hundreds of minor celebrities compared to these greats, Syed Hashim Raza, AT Naqvi, Sikandar Mirza, writers, teachers, scholars and politic taints, who created Pakistan, in which they are now alien and mercilessly killed each day.

On this February gray and cold afternoon, standing opposite this painting of Teepu Sultan, I just wanted to add one more name to the illustrious list—that of Sultan Fateh Ali Teepu, son of Sultan Hyder Ali and Fatima, Fakhrulnisa of Mysore, who could easily eclipse all those in the list of Shias who fought for the liberation of Muslims from outside invaders over centuries.

As one famous anchor on an independent TV channel pinpointed, the Arab money is being funnelled into Pakistan to eliminate the Shias from this land, which was in fact created by their forefathers!

Earlier on, in 1492, three fateful events took place in the history of world. Moorish Spain was lost to Richard and Isabella of Aragon; and the last of the remaining Moorish Kings in Granada had to escape for his life to Morocco. Thus ended eight centuries of Muslim Spain. What began as an adventurous gain by Tariq bin Ziad and Musa bin Naseer, when they fought and climbed the cliffs of Gibraltar alias Jabal-e Tariq, a stuff of legendary tales, ended up in *mahalati sazishen* (Palatial intrigues) and disaster. Andalusia now cries for the sound of an Azaan! Iqbal lamented so in his famous poem, called the *Masjide Qartaba*.

The second event that happened the same watershed year was that the Diaspora of the Jews began. They had prospered under the Moorish kings and produced philosophers and physicians like Maimonides but had to flee for their life under the

Christian persecution and what is now called ethnic cleansing (after the Bosnia Herzegovina events of the recent times). In order to be economically independent, they decided to take up sciences and professions like finance and economics, medicine and education. The rest is history.

The third and perhaps the most relevant event to contemporary times that happened in 1492 was the case of mistaken identity, not of person but of a place, by a seaman called Christopher Columbus. He landed on the soil of America thinking that he had finally found the lucrative route to the near and far East, to the lands of silk, myrrh, spices and gold. Such was the glory of India and its neighbouring states that it lured the seafarers like Columbus to find and exploit the treasures of those lands.

America was discovered and has since dominated our lives.

Around that time the name of a man called Alberkerque crops up in the annals of the Spanish wars against the Moors as the general who spearheaded the attacks. His name also appears in Manilla, where a few years later either he or his descendent by the same surname fought battles against the Muslim king of the Philippines, called Suleiman. The king was defeated by the general and jailed in a dungeon along with hundreds of his loyal subjects, who were then drowned by the river at the bank of which this dungeon was made. On a trip to Manilla, one can see this dungeon as it sits in close proximity to the river that was used to flood the dungeons. One can actually see the floodgates that exist as the dark reminder of those fateful events and the bravery of King Soleman and a national hero called Rizal, who gallantly walked to the gallows but did not surrender to the French and Spanish invaders. Rizal's footsteps to the gallows echo forever in memory, and people cherish them.

Though the Muslims lost an empire in Andalusia, by then the Ottoman Empire had conquered most of the lands in the near and the Middle East and had put up a siege upon Vienna. Constantinople was conquered by Sultan Mehmet in the 1480s, and Hedja Sophia's cathedral had become a joint place of worship for both Muslims and Christians by 1482. Muslims had also ventured into India and gained huge territories in that land of the 'infidels' also. All those stories of the adventures of the Turkish sultans that we were taught of at schools bring us back to the present day Pakistan, for it was this very land that their horses had to trample upon before reaching their desired goal, i.e., the wealth that India possessed. And that attracted to India the European seafarers like Vasco De Gama, who landed on the coastal areas of southern India in 1498. Not long after that, European traders gained inroads into India. Once again, let me say that the rest is history.

The 'land of the pure' is blessed with natural barriers in the form of mighty mountains and deep seas. In the north stand tall the Hindu Kash and the seven glorious peaks of the Himalayan Mountains, including the K-2 and the Nanga Parbat. In the south the deepest of the oceans called the Arabian Sea. In the East is the ferocious and ruthless desert of the Babylonian proportions called the '*Thar ka sahra.*' And in the West the harsh and ferocious rocks of Baluchistan. Even the names are symbolic, such as the Kohe Murdar!

These natural frontiers safeguard the present day Pakistan and have done so for several millennia. The only natural routes of entry are through the famous passes, namely the Khyber, Gomel, To chi, Bolan and Khurram—or indeed, through the mighty spread of sea water only ventured upon by Mohammad Bin Qasim et al. The final route of entry is through the merciless sahra of Thar, which actually spread out into the vastness of the deserts of Jaislamir, and beyond that into ancient Mesopotamia, and Nainawa, imperceptibly merging into the emptiness of the sub-Saharan Africa and its deserts.

Only an unwise or perhaps ill-advised king took the desert route, and that was the ill-fated Humayun. He went to Iran looking for help, via Sindh, and camped at a place near Khairpur called Baburlo, or Babarloi, meaning Babar's camps, or tents. Its date palms, called *Khaji*, are exported to the California supermarkets.

India, and therefore Pakistan, have benefited as well as suffered at the hands of outsiders. For a long time, the ancient route of the Khyber Pass has been the safe haven for the passage of hordes of invaders from central Asia. In the so-called Minor Asia, the lands were mostly infertile and the weather inclement if not outright merciless. Tiny kingdoms had sprung up hither and thither, fighting for their survival. Their meagre means and their skills no more than those of riding a saddle-less horse and sabre rattling, these tin-pot soldiers found their ways from the rooftop of the world called the Palmirs down the snow-clad hills and valleys towards Kazakhstan, Balkh, Bokhara, Konia Doshanbe, Mazare Sharif, Jalalabad, Kabul, Peshawar, into the fertile lands of the Punjab. While the generals were busy conquering the world, the saints and dervishes were spreading education and promoting the welfare of mankind, through no lesser persons than Shams Tabriz, Maulana Rom, and Imam Bokhari, and above all Abdullah Shah Ghazi of Clifton, Karachi. He is the first grandson of Prophet SA to arrive in India, an *Imamzadah*, who regrettably is not given his due status by the hypocrite nation living around him.

Earlier on, the Turks like Al-Patgin, Subukatageen and their following dynasties including their slaves and their descendants had ventured into India, and reaped the harvests for their greed and huger. Of course all in the name of Islam! These dynasties had no intentions of settling down in India, as they missed their open-air lifestyle and some, with the nomadic background, the life under a clear blue sky, star gazing and all those romantic preoccupations that our folklore continues to remind us. Despite his lavish palaces and so forth, the ill-fated dictator of Libya, Colonel Ghadafi liked to enjoy his tent and entertainment of sorts under the sky albeit within the air conditioned and luxurious tent, pitched out in the desert of Tarablis. His disgraceful departure in 2012 as a part of the so-called the Arab spring is a living proof of the fate that all tyrants meet. *Ya Oolaial Absar*!

The whirling dervishes of Constantinople and today's Istanbul were duly introduced to India in the form of the fakirs whirling away on the rhythmical music of Sain Sachal Sar Mast and Lal *Shahbaz qalanadar* in the shape of *Dahammal*. So some things have stood the test of time and continue to remain a great source of pleasure to us.

Babar had to escape his tyrant uncle and greedy cousins. Being a cunning man he escaped the family feuds and fought his way into India. His legendary bravery is told even today, that he rode a saddleless horse and swam across the width of Kabul River and so forth. Anyway he came, saw and liked it; hence, he conquered India.

He, however, always wanted to return home, which even today is called the '*muluk*.' He preferred his watery watermelons to the gouvy mangoes of India! (If only Gahlib was there to reprimand him!)

The love of the '*muluk*' is so strong that despite living all their lives in Karachi, most of the Pushtoons prefer to retire to their '*muluk*' after earning their superannuation. Such is the might of the love of home soil!

All kings, invaders, warriors, adventurers, and fortune hunters entered through the Khyber Pass and winding their way down the eastern banks of Indus, they passed through the hard and sturdy territories unto Lahore. Some say that since their horses could not cross the ferocious river Indus at a place called Attock, they preferred to follow the natural route towards Delhi and not venture into the southern territories of present day Pakistan. One wonders if the word Attock is self-explanatory in a bizarre way.

Bypassing Lahore or using it only as a temporary cantonment area, the armies marched on through the green pastures of east Punjab and finally reached the destination of all kings of the past and present, namely Delhi. They would happily settle down in Delhi and Agra and then control the distant lands of far and wider reaches of India with forceful authority. However, they all sent their booties back home via the time-trodden route of the Delhi, Lahore, Peshawar, Khyber Pass, Kabul, JalalAbad, Mazare Shareef, and into Asia manor. Right through history, from Mohammad Bin Qasim through the early Arabs, the Turks, the Selduiks and the Mughals, India was a place to earn good fortunes, but not hospitable enough to permanently settle down in. Even the famous warrior Nadir Shah came to pay a short visit to enrich himself. So did Ahmad Shah Abdali and his many cronies. Earlier on, Mahmud of Ghazni was obliged to pay several visits depending on his need, as well as greed. There was much to loot, and he had ample time to return to rob!

Mushtaq Yousefi, in his humorous style, recalls how many dates of the invasion of Somnath did he have to memorise in his history lessons, and questions as to why the man from Ghazni could not rob India wholeheartedly in one ambush only, thus saving him the ordeal of recalling multiple dates.

Akbar, the clever king, was a visionary in many ways. He realized quite early in life, partly thanks to his illustrious *ataleeq* Khan e Khanan, that the land of Indus offered much more than the inhospitable hinterlands of Fargahana. Hence Akbar decided to settle down for good in Delhi and Agra, leaving the time honoured trail of Lahore Jalabad, etc., for other adventurers like the later day Taliban, etc. He enjoyed the highly developed culture and civilization of the valleys of Ganaga Jamuna, alias wadi-e-Gango Jaman. He left the discomfort of a saddleless horse back for the comfort of the palaces, even building a city called Akbar Abad, to suit his own likes and pleasures.

Though illiterate, Akbar was a true lover of arts and culture. He hunted down the finest talent in art and music and culture and theology, even inventing his own religion. He enjoyed the prostration and obescence of the French ambassador to his court so much that he ordered all his courtiers to follow suit and prostrate in his presence when he bestowed the honour of his presence. The word Courniche is the outcome of this practice. Sycophants have always grown in every culture like fungus in dampness; hence, noticing that the King Emperor likened himself to be godlike, they started calling him Zille Elahi, i.e., the 'Shadow of God.' Akbar was no fool. He exploited the situation to his egoistic satisfaction even more and went one step further. He commanded that anyone entering his courts should shout, 'Allah Akbar,' a normal and even desirable slogan for a Muslim, but not so for Hindus who formed the bulk of his subjects. I suspect even they had to raise the slogan to please Akbar, who cleverly disguised his own desire to be god behind this common slogan, 'Allah is great,' also implying clandestinely that Akbar is god!

Since Akbar also realised that he had to win the hearts and minds of his multi-faith subjects, he evolved a new form of salutation and replaced 'Salaam -O–Alaik' with 'Adaab'; i.e., respects, a saving grace for the nonbelievers. Until today, this form of salutation is practiced when a Muslim and a non-Muslim meet and greet each other. Such was the genius of Akbar in amalgamating different civilisations into one homogenous society.

He also recognised that dance is an art which was part and parcel of Hindu culture, more a religious custom than a matter of plain joy. He developed it further and built up institutions to promote and refine the Indian dances into perfect forms that we continue to enjoy today. His mountainous ancestors probably enjoyed and participated in male-only dances, perhaps some form of the present-day Khatak dance seen on the festive occasions in the tribal areas of Pakistan, with sabres shining and rattling; or sometimes just to tone it down for the weaker hearts, colourful handkerchiefs briskly and artfully waved across the face of the opponent by the dancer.

Akbar probably helped in converting that rather wild form of dance into a subtle and soft poetic form of movement of hands, feet, body and torso compounded with a pretty face of a damsel narrating a whole love song through what is called a Kathak nauch. So the Khatak dance was deftly eased down into Katah kali and its cousins like Bharat natayam, and munipurri, etc., by those fantastic visionaries of earlier days.

Dhol or the drum is still seen making horrendous noise in certain weddings in the Punjab and elsewhere. Amir Khusru, a genius at music, realised that the harsh and crude noise of the Dhol was often enough traumatising to the nerves and did not make much sense; nevertheless, it was an essential accompaniment of the orchestra, so he divided it into two halves and called them Tablas, a word derived from Tabal as was Tabl-e-Jang or the drum beaten at war.

Religious music is a part of Hindu and Sikh faiths, but not allowed in Islam. Most of the emperors thoroughly enjoyed the luxuries of life, so how could they give up the art of music particularly during the religious months like Ramadan; hence, a combination of traditional and classical music was specially developed in a religious format to please the king and the courtiers. Just like the Bhajan for Hindus, the Naatia Qawali was developed for the Muslims. It then filtered down to the masses as it is customary for a commoner to do what the king and the courtiers do. Of course, Qawali has since grown into a global phenomenon, thanks to legends like Ustad Nusrat Fatheh Ali Khan and many others.

The dancing girl of the ancient times of Moen jo Daro, a legend in its own right, was gracefully developed into an artist par excellence and the latter day of the Oudh dynasties of Wajid Ali Shah et al. cultivated it even further to make it a fairly honourable if not entirely respectable profession. It is said that the rich and the famous used to despatch their offspring for learning the ropes of art and etiquette at the feet of these courtesans and their Ustads.

Henna was used forever in Asia minor, but more for dying the beard to look less than your age, than for deftly decorating the delicate palm and feet of maidens, as was the practice in India and continues to remain so until today. Henna tattooing is now catching up in modern Britain also.

The Betal nut chewing has always been a curse in Indo Pak subcontinent. Its variant in the form of Kat chewing is seen in Yemen, where it is almost a national symbol. The dynasties in India converted the betal nut chewing into a dainty and popular pastime, developing many forms of perfumed and aromatic forms of chewing tobaccao called Qiwam, adding cardamoms, cloves and many other ingredients. A habit that has stood the test of time and despite adverse reports and universal condemnation, a Khushboo ka paan is a luxury that few can resist after a sumptuous dinner at a five-star hotel in Karachi, Lahore or Islamabad. But not so in Peshawar or Quetta where it is duly replaced with a different form of chewable tobacco called Niswar, which has more in common with Kat than Paan.

The mughal army was huge, and had different tastes for food. Some ate only meat, others only rice and lentils. It must have been a nightmare for the logistic department dealing with food and rations to cook so many different meals for an army that always remained on the move and obviously starved. Therefore a genius chef saw an outlet and saved himself and his generations of the tedious job of catering to a wide array of taste buds. He took a huge cauldron and kept dumping all that he could get from the logistic department; i.e., meat, and daal, and wheat and salt and chillies, and onions, and nutmeg and a whole lot of other ingredients to emulsify the fat and proteolate the meat fibre, cooked it on a slow heat and fed the army to its satisfaction. That dish is called Haleem, so fondly consumed by many a people in Indo Pakistan even today.

The courtiers of the Mughal courts had many different types of attire to wear. It was all very colourful and full of variety. The king emperor continued to wear the loose tunic almost as did the ancestors of the bygone days. The clever courtiers realised that some Hindus felt left out with the Arabic dress code of wearing a loose abaya, and some Muslim courtiers envied the grace and formality of the buttoned-up collar of the Rajputani coats. The clever tailors of the time amalgamated the grace and the flow of the Arabic abaya with the formal outlook of the Rajasathani coat and developed a dress called Sherwani, so formal yet so suitable for the warmer climates of the Indo Pak regions. Its cousin, called Achkan, became a legend in its own right at Aligarh, which became the cradle for the evolution of Pakistan in the earlier parts of the twentieth century.

Shalwar is primarily a modified version of pantaloons worn by the Turks and mainly the Kurds. The whirling dervishes harnessed the ample enclosure of their Shalwar for gathering sufficient air during their volatile whirl, to literally lift them off the floor. It was not a popular dress in UP where Aligarh kat trousers, called Khara pyjamas, were and continue to remain the common dress. Today's Indian films are a witness to this statement. In Pakistan the migrant population from UP found the Shalwar a baggy and rather unworthy dress to wear in day-to-day matters. It was considered to be more suitable as a comfort dress than a formal one. But as the time passed by, all these muhajirs gradually became influenced by the Sindhi and Punjabi culture and the shalwar all but replaced the Aligarh Kat pyjama. Such is the influence of culture of the dress code!

In modern Pakistani dramas, a ferocious-looking *malik, wadera or a sain* often represents the true nature of the male-dominated society as a tyrant, and a ruthless zamindar. This hero usually wears a heavily starched shalwar kameez, a heavy beard and matching moustache. The whole persona is designed to inflict fear in the heart of the onlooker.

If only such folks had ever read about the tyrant of all tyrants who terrorised the whole world in his times. He was still to come.

A wise man called Luqman had advised his son not to walk haughtily on earth as he could not break it, or indeed touch the skies above (despite his Kulah, that Ghalib referred to in his famous couplet referring to '*darazie qad'*), nor to raise the voice either, as the worst of the sounds was the braying of an ass. But who cares about Luqman or the word of God?

Enter the mighty Jengis Khan, the wholehearted warrior and a tyrant. He killed a boar or a bear to make his hat, which was warm, and saved his majestic though illiterate head from soaking in the wintry rains of Mongolia and the Gobi desert. The man has bestowed us with a fine headgear. His hat continued to be transferred through generations into the various forms of caps we wear. The softer the personality and warmer the climate, the softer and lighter the hat. Hence the Qaraquli continues to remain popular in the cold climates close to Afghanistan, but the natives made a hat called Gandhi toupee, which rose to a popular place in the warmer regions of Gujarat, the central and western India.

Indeed there are two distinct versions of Islam that we see in Indo Pakistan: the rough and tough variety of puritanical Islam in those areas demarcated north of Marghalla pass quoted by the National Geographic, and the softer version of Islam in the rich and fertile valleys of Sindh and Punjab. The Sufis and saints, who often enough escaped the tyrannies of men like Hajaj bin Yousef and their hatchet men to hide in the remotest corners of India, particularly Punjab and Sindh, are a living proof of their service to the cause of Islam. While the tyrants killed and maimed the people in the name of religion, these Sufis and saints harnessed the ordinary folk around and taught them the peaceful aspect of Islam.

Hanif Ramay was a renowned politician, writer and an artist. In his recent book he described that all those warriors and conquerors that came to India did not particularly endeavour to convert Hindus to Islam, as it was a good source of constant revenue in the shape of Jizzya or religious tax. So for obvious reasons they preferred to maintain the status quo, albeit Islam spread through the good works of the Sufis and dervishes. Aurangzeb was a religious zealot, a fanatic warrior, a bigot and a tyrant. He imposed the jizya after Akbar had abolished it years ago and spent

16 years of his life and times in South, trying to abolish the Deccani Qutub Shahi states, only to face the onslaught of Marhattas, who eventually destroyed him. According to Muni Lal, a famous Indian historian, Aurangzeb hated the Shias so much that he named his personal dagger as 'Rafzi Kush.' Such are the heroes of Islam, one is ashamed to say.

William Dalrymple, a wonderful contemporary writer in his recent book called the 'Last Mughal,' mentions it in his description of the lamentable decline of Bahadur Shah Zafar, that towards its terminal days the royalty was so deficient in cash that one of the princes gave strict orders to put all the 'kafirs' under excessive pressure to pay the jizya (so that the agas of the battle could be kept aflame)!

The cultural variations are so enormous within the boundaries of Pakistan itself let alone the region that one cannot generalise or accept a universal code of conduct without at least modifying it to suit the local traditions.

In a BBC documentary shown on BBC TV on August 14, 2007, a Pakistani lady, broadcaster, showed the panoramic colours of the excellent and often breathtaking scenery of Pakistan and the habits and customs of its people. The colourful trucks across the length and breadth of Pakistan traversing the highways and byways adds so much attraction and colour to otherwise monotonous and rather bland scenery , that by itself it has acquired a title of truck art in Pakistan. It is loved by the British. At an Asian fair in Birmingham a couple of years ago, someone had brought in a Pakistani bus, painted in its yellow and orange and green and blue colours, with numerous motifs written on it including the one so popular; i.e., '*Papoo yar tang na kar.*' The fete would have all but failed had it not been due to this humble Pakistani bus that attracted throngs of people to pose in front of it for portraits in all kinds of postures.

The Makarani donkey cart in Karachi has its humble status, but if you speak to the proud owner of such a cart, he'd call it his Rolls Royce or Daimler and caress it with love and cherish a ride on it in the evenings in Lyari, boasting about the speed of his pair of donkeys. It is also a part of their tradition that of the two donkeys, only one actually pulls the wagon; the other called the '*Pukh*' is employed just to keep the main donkey perpetually challenged lest he be overtaken by the other one! It speaks volumes on the intellect of the ass.

The Makranis who inhabit parts of Balochistan and Sindh are a very special race. They are mainly concentrated in Lyari area of Karachi and until a few years ago, when the cinema houses were still populated, you could see them maintaining order in the queue for a ticket with their menacing although benign vocal threats.

They are called Sheedis. Interestingly enough, in most of sub-Saharan Africa, the common terminology of Sidi to address a gentleman is a popular practice. Now, if we trace the roots of Pakistani Sheedis or Sidis, one may be able to trace them to the lands of Tanzania.

Until 1958 or thereabouts Gwadar was a part of the sultanate of Muscat. Its Sultan was also the Sultan of Tanzania Darul Salam for the last few centuries. Even today two forms of Muscatis are known to inhabit the lands, the Zanzibaris and the Omani Arabs, both with their distinct cultures. So it is a feasible theory that our Sheedis or Sidis were brought in by the Sultan of Muscat from Tanazania to serve his majesty and his subordinates, perhaps as unpaid workers alias slaves.

Their customs and traditions are very much like their racial cousins in Africa. An interesting observation is that the black community across the globe is very active. Not only are they physically strong and have superb physical features, but they are also very mobile. You would never see a march, even if it is supposed to be a sombre occasion or a mark of protest anywhere in the black world where people would walk sedately. They must move in rhythm, singing and dancing, having a great time, without a care in the world. Even the finest of cricket players like Vivian Richard would dance to catch a sky-high catch rather than desperately run for it as some other lesser mortals would do. The black church is the finest place to go to. They sing and dance and make merry. God must surely be happy with their own distinctive way of worship! Another facet of cultural identity.

Nelson Mandela now ailing, is a symbol of valour, honour, fortitude and dignity, and these are indeed only some of the many virtues that this living legend possesses. If you note his body language, despite his age and frailty, he still shows the bodily rhythm of walking that used to be a dance until a few years ago. So the African race is a happy race, and despite the history of the hardships they went through in the sixteenth and seventeenth centuries, they have so much dignity and self-esteem that one has to admire their bravado.

The Sheedis of Lyari have such typical values and customs that one can hardly generalise them with the rest of the populations in Pakistan. Each tradition is built upon mighty strong historical facts and covers a vast plain of geographical dimensions. One only has to see the Makranis at the Mela of Pir Lahoot Lamakan to understand their distinctive cultural identity, which has little in common with any other inhabitant of the land.

Many years ago a renowned professor of psychiatry in Karachi filmed them during a religious mela, when they went into a state of trance and began to speak Swahili and other African songs of praise, which they did not ordinarily speak or understand.

Pakistan is a land of many diverse cultures but one religion. If religion alone was the binding force, then the entire ummah should belong to one nation, one country! And that is perhaps what Allama Iqbal meant when he claimed, "*saray jahan hamara*"; however, he had to hastily add, "*Muslim hein ham watan hein Hindustan hamara*." Or the Arabs and the Turks would get offended, as they were accustomed to invading others rather than being subjects of surrender to other nations, even if it be hypothetical.

No doubt the code of conduct for every Muslim across the globe is universal, such as the rituals of Salat and Zakat, the burial and birth, etc., but cultural impact on matters of ethics and bioethics are simply too strong not to be taken into account. Each country has its own culture, which has a deep impact on their ethical values or at least how they practice them.

Pakistan has a strong Bradari system, which is seen in many other brave and valiant nations. Remember the Red Indians, who had such a strong brotherhood and a powerful Baradari, They fought and died honorably, than accept slavery. The slave traders had to trap Africans of different countries, to catch them and transport as slaves to the American colonies. Bristol, Liverpool and Southampton are called the slave cities, as many slave ships sailed to America from here. In Liverpool, not far from the docks just opposite to the immigration offices called the India House, there are several ancient buildings displaying Negro slaves in collars, hands and feet shackled and worse, etched out on the stones that builders used, so as to eternally display the might of the British and the slavery of the black races! How woeful.

In his autobiography, General Parvez Musharaf quite rightly points out that he was given the treatment that he was given by Nawaz Shareef as the Prime minister, as Shareef knew that Musharaf was son of an immigrant and did not have a clan to support or to lament upon him. Such is the might of tribalism in Pakistan. Musharaf had no clan, so he lived in exile and Shareefs had huge Bradari to support them. So they are back in power.

The only other head of Pakistani government, besides Musharraf, to face such a lonesome exile was Iskandar Mirza, who also did not have a clan in Pakistan or elsewhere. Allegedly he served as a manager in a famous Indian restaurant off Regent Street in London, to save himself from going on benefits! And yet many cowardly generals, presidents and prime ministers who had their Bradari to support them continue to live on the soil they left, no stone unturned to sell, plunder or surrender to other nations.

At the demise of East Pakistan and the loss of nation's dignity as indeed suffering humiliation and disgrace, the reigning monarchs of the broken and beaten-up Pakistan devised a strategy to masquerade their blood-stained daggers under a cloak of disinformation.

Eminent scholars were hired to discuss and debate, as indeed to promote on the national media that the loss of East Pakistan was not the end of the world, as Pakistan could now take up the leadership of the Arab world as the roots of the residents of residual Pakistan stretched deeper into the Middle East than in South East Asia. The implication is, of course, that those dark-coloured, rather small-framed Bengalis had more in common with those in India and South East. Now that they are no more there, we should go back to our Middle Eastern roots!

One reason of course was that Dubai had just discovered massive quantities of oil and the rules knew little more than the care and maintenance of their wooden dhows. So Pakistan with its brainy bankers like Agha Hassan Abidi stepped in to manage the massive funds generated by the oil production in these crucial states.

The attempt to totally disengage the South Asian roots and celebrate the Arabo Iranian heritage did not catch up with the public. Besides, Arya Meher, the mighty king of Iran, was waiting in the wings to lead the oil-rich Middle East anyway. Why indeed should he allow the poor albeit brainy Pakistanis to become their leaders!

Therefore, no decision could be reached as to the real culture of Pakistan despite the massive efforts of legendary scholars, and the dilemma continues *ad infinitum*.

Enter Ziaul Haq, the self proclaimed 'marade momin' of sorts. He attempted to falsify the fundamental philosophy of creation of Pakistan from a saccular land mainly for Muslims to a new slogan. It was '*Pakistan ka matlab kia, LaIllah Illal Lah.*'

Surely Mr. Jinnah would have reprimanded him if only he were alive today. Zia led to Talibanisation of Pakistan and destroyed the very fabric of the Pakistani society. It was the hatred implanted, sowed and nurtured by him that led to fragmentation of the society on sectarian basis. Many a thousand were killed or banished from the land of the pure to impure lands for their religious beliefs. He portrayed himself as Aurangzeb, as against Bhuttos, who allegedly represented Dara Shikoh!

The dilemma continues for a nation simmering with the aftermath of the capture and death of the founder of Al- Qaeda, only a few minutes walk away from the Sandhhurst of Pakistan!

Those who love Pakistan and live in exile, self-imposed or forced, lament at the decline and fall of a country that could have been a front leader had it been allowed to prosper. The cast for its demolition was made with the murder of a genuine, honest and dignified leader, who was baseless in the land that he represented as its prime minister. Liaqat Ali Khan came to realise soon after migration that he could not win a parliamentary seat on his own, as he was not the son of the soil. Same formula would replicate years later in the forced departure of Musharraf.

Shabeeh ul Hasan is an eminent scholar, thinker, writer, and a cosmologist. He points out, in an essay called the *Pakistan ka myth*, that in 1946 the Delhi resolution clearly identified and earmarked Bengal Punjab, NWFP, now Pakhtoon Khwa, Sindh and Balochistan to form Pakistan. So why, he raises the question, did the leaders entice, coax, malign, even coerce, all the other Musalmans who lived in UP, CP, Behar and elsewhere, to join the movement, if they were not to be a part of Pakistan. He concludes that all of them left their home and hearth to end up in a '*bund gali,*' an abyss. Liaquat Ali Khan was one such person, one suspects.

On the subject of Indo-Pakistan, Syed Shabeeh ul Hasan, in his book called *Nairang e Haqeeqat* simply excels in expressing his views that may represent many millions of those who suffered in and after that fateful August of 1947. He strongly criticises the 'two-nation theory,' which formed the basis of the division of India.

He is particularly critical of the dream of Allama Iqbal. According to Syed Sahib, the dream of the poet-philosopher envisaged creation of the new state specifically in areas in the northwestern regions of India. Those areas were already under the Muslim control. Allama did not include the rest of the Muslim populations living in India in his visionary Muslim state. Shabeeh sahib raises the question, why did the politicians then agitate the Bengali Muslims and those settled for a thousand years in UP, CP or elsewhere? Allama did not envision a state for these folks, so why displace them? Besides, no consideration was given to the island states like Hyderabad and a thousand small and large dukedoms. Was Allama Iqbal's dream distorted to suit certain vested interests? Shabeeh Bhai has tackled this point philosophically and historically, tracing the roots of the troubles, to the times of the first Arab invasion of India.

His point is duly emphasised when one looks at the plight of the so-called Mohajirs, who left their home and hearth for the new 'lands of the pure.' They became homeless and baseless forever. Why, he asks, did the leaders misguide them or not take active measures to avoid displacement of humongous populations, bringing them untold miseries? History has not seen a Diaspora of such magnitude ever before, not even during the Biblical exodus of Bani Israel from the Kanaan valley.

An interesting viewpoint that emerges out of the discussion about the future solution of the current dilemma of the subcontinent is the formation of a larger federation akin to the United States of America. It may include India, Pakistan and present Bangladesh, he argues. An eminent visionary of the past Maulana Mohammad Ali Jauhar once wrote an editorial in his paper called *Comrade*, where he proposed a solution on the similar lines. Alas petty politicians did not pay a heed to possibly the best solution of the Indian problem after the British left.

Tariq Ali, a well-known leftist political historian- journalist, now living in the UK, wrote in his book called the *Protocols of Elders of Sodom* that after the separation of Bengal; i.e., East Pakistan, the remaining Pakistan is "no more than an extension of Punjab, as the other three provinces are sparsely populated." That certainly seems to be the general opinion of Pakistanis, locally as well as those spread out in the shape of the Diaspora across the globe, that Pakistan is indeed Punjab, the others just happen to be in the neighbourhood! A former eminent politician, now dead, described it as triangle, in which Punjab sits at the top of the triangle as the army, on one side of the triangle as the bureaucracy and on the third side as the industrialists or *a malik*. In most influential Punjabi families, one brother joins the army, the second one joins the civil services, and the third stays on the lands (sometimes also practicing law from a circumspect law college just in case!). The other nationalities are encircled by Punjab to live at their mercy, if you like.

If Pakistan is nothing but Punjab, then why indeed did others have to suffer in 1947 and continue to do so even now?

Another question which therefore crops up is 'Should Punjabi culture be accepted as the true representative culture of Pakistan?'

The saga of Pakistani soap continues to baffle everyone. Only last week someone at a Faiz memorial seminar held in Birmingham, said that when Faiz was asked to predict the fate of Pakistan as he returned after his Beirut sojourn, the formidable poet reportedly said , *"na tootay ga na sudhray ga , kamabakaht aisay hi chalta rahey ga!"* What a prophesy from a revolutionary thinker indeed!

Therefore, to sum it up, one cannot dutifully, honestly, and clearly define the culture of Pakistan. It does not appear to be a nation of homogenous people but a conglomeration of many different ethnic groups, who like a bad marriage are resigned to the circumstances, and perhaps in Samuel Becket's words waiting for Godot (or a messiah).

With that information on cultural anthropology, we will now study the rules of Indo-Muslim bioethics. We will also look at some models that might be of interest as part of this discussion. Learned ethicists Beauchamp and Childress have identified four pillars; we would like to evaluate them on the fundamental rules of ethics given in the dictates of Quran, namely Al *Takreem, Al Birr-Wal Adl Wal Ehsaan*.

4.2   Normative Principles of Medical Ethics

4.2.1   *Al Takreem Al Birr, Wal Adl Wal Ehsaan*/Autonomy, Justice and Beneficence

All faiths and religions have taught their followers to respect human dignity and freedom. In fact, it is not necessary to have any form of faith to uphold these principles. In today's faithless world, there are arguably more nonbelievers doing good deeds than the believers who are often misguided by the illiterate mullah.

Each culture has its own tools to measure the dimensions of human values. No doubt there are some founding principles that are universal, such as human dignity, individual's rights or the rights of the community, but autonomy per se may have different implications in different cultures. Human dignity is indeed the fundamental rule and the guiding principle behind the concept of informed consent, as advocated by both through Western greats like Socrates, Aristotle and Plato, as well as all the religious faiths. Islam attaches singular importance of the rights of an individual, the weak and the downtrodden, the women and the children, the orphans and the destitute.

In his Hajj sermon, Prophet Mohammad SA gave a lasting verdict, which is the first document of its kind. It is universal in its approach to all mankind. Its message is quite simple. It is based upon the principle of human dignity, and the rights of an individual irrespective of his colour, creed, race, or the position in a community. Holy Prophet, in fact, drew out a road map for the whole world to follow and respect everyone without prejudice of any kind. He commanded us to grant freedom to mankind from all kinds of exploitations, e.g., slavery, usury, malice, avarice, envies, or jealousies. No one, he said, is superior to his fellow being except on the basis of *Taqwah*. Allah grants closeness only on the basis of piety and piety only; all other qualifications, achievements and attributes, however glorified they may be, carry no meaning with Allah. Islamic ethics is derived from

1. the noble Quran
2. Sunnah
3. Ijma, Aql o Idrak
4. Ijtehad

His closest disciple, Ali Ibne Abi Talib, said in one of his sermons, "You must respect each and every human being as either his your brother in faith or brother on account of his birth, as everyone is a son or daughter of Adam and Eve."

In the eyes of Islam, *Takreem;* i.e., dignity, enjoys the top position amongst all other worldly matters. Any form of disrespect or indignity to an individual or the society is categorically condemned by the Quran. Islam discouraged slavery, and duly commanded its followers to free the slaves long before Abraham Lincoln's freedom-of-slaves declaration.

Qanbar was Ali's devoted servant, who was encouraged by the master to dress better than the master himself. Such examples are eye-openers to any spokesman for the human rights. Fizza, the slave girl of Fatima al Zahra, was addressed as mother by Hasan and Hussain and respected just as much. Fatima shared the domestic chores with Fizza on daily basis. There are lots of similar stories that come across the annals of Islamic history. Unfortunately, with the advent of *malokiat*, the guiding principles of the Prophet and his family and the immediate followers were duly forgotten by the Syrian kings and their descendants.

Ancient Greece, much before the Islamic era, is considered to be the cradle of the philosophy of human freedom and human rights. Some of us may read the story of the Sparta and Spartacus, where slavery was a norm. In fact, mighty philosophers of ancient Greece do not strongly condemn slavery; on the contrary, Aristotle is known for his bias against certain classes.

In modern Greece, the land of the great philosophers, the Western model of autonomy is no more an accepted norm. Individual autonomy is neither practiced nor promoted in today's Greece. The physician has a strong position in decision making, and is granted the solemn status of a senior and a wise man who knows it all. He is taken as a father-figure, a teacher, a thorough professional, who could do nothing wrong, and would always give priority to the patient over other matters. The paternalistic approach in today's Greece is almost akin to what we see in most Asian countries.

Some researchers have established that the Greeks often prefer that the physician would not tell the whole truth to the patient as it may bring more harm than good to an individual or the family.

The similarity here between what is seen in the Indo-Pak subcontinent and modern Greece is startling. In Pakistan, also, the classical paternalistic approach is seen even today, though in the cities some changes are beginning to show. Pakistan, having a single faith but multicultural society, has many colours of autonomy to witness in its various regions.

Islam is the major religion in Pakistan, but its impact is significantly diluted in day-to-day practices by the powerful cultural influences. What applied to a Baloch tribal leader cannot be replicated to a commoner in the streets of Lahore or Karachi, even though they both are taught Quran and are fully aware of the last sermon of the holy Prophet.

In clinical practice also we see different models across the country. Education and enlightenment are major differences between different models. Nearly all physicians, though, enjoy tremendous status in the society. Paternalistic approach is almost a uniform principle in most regions: the doctor knows best and we put our faith in Allah and after Him in the physician. It is an oft-repeated, oft-heard statement anywhere in the country. Just like the Modern Greek model, the physician is often asked by the family not to tell the whole truth to the patient as it may precipitate his condition. It is particularly the case if it happens to be malignant disease.

In a case of terminal illness, avoiding the news of the inevitable is often encouraged by the next of kin. Perhaps in many ways it is kinder too, as sometimes withholding truth may be ethical, particularly if one looks at the possibility of bringing harm more than good in such an event. The fact that certain disclosures may bring about instant harm or cause distress, bring about anxiety or stress to the near and dear ones may compromise the principle of autonomy to a certain extent, but since the ultimate objective is Beneficence, avoiding malificence to the patient, it is justified in our society, which is closely knitted like the finest muslin from Dhaka. You remove one tiny string from that fabric and may end up with a big hole in it. Such a strong family bond is almost nonexistent in the West, and therefore it is simply not possible to blindly follow the principle of autonomy and informed consent without taking the cultural variations in view. Besides, in such situations, the communication experts dealing with problems like breaking bad news, etc., inform us that often enough it is wiser to grant due status to mercy over commitment to duty alone. Indeed mercy is a far superior virtue than fulfilling one's dutiful obligations.

As mentioned above, in modern Greece, a thin line divides truth-telling from nondisclosure. No hard and fast rule governs the population at large. In the rural areas, just like in Pakistan, paternalistic approach is the standard practice; however in the cities, the educated patient likes to know everything about the illness as well as the remedies advised. The same is the case in cities like Karachi, Islamabad, Lahore, etc. In fact, it is a delight to be able to discuss the problem with the patient and be able to jointly take the decision in matters of medical care, particularly intervention, wherever possible.

In the land of Socrates and Pythagoras and of many sophists, before him, by and large the principle of beneficence and nonmalificence is duly observed rather strictly. They therefore genuinely feel that the patient need not know everything all the time. Such is the mutual trust that binds the physician and the patient in a state of harmony, understanding and best interest of the patient.

In the Islamic world, faith in Allah as the supreme power plays significant role in all matters of life and sickness. Faith and trust unto God is followed by trust unto the physician, who, as most patients believe, would treat his patient keeping in view that that he is under the constant vigil of the Supreme Being. Health and sickness both come from Allah and we have to meet our destiny one day. Each one has to eventually return to the Creator. Many times, in matters of tragedies and terminal illnesses, the guiding rule is a major source of solace to the sufferers. Sickness in many ways is a trail of an individual and the family, and God alone in His infinite mercy grants the necessary courage to bear it. Freedom from illness and return to health is gift from God Almighty; therefore, a salat of gratitude *Namaze Shukrana* should be offered to the creator for saving one from catastrophe and returning to health and the joys of life.

In Pakistan, just like in other Muslim countries both Arab and non-Arab, prayer for the recovery of the patient and a *Namaze Shukrana on* achieving the recovery are a standard practice.

In Japan, which is a model for discussion as a highly industrialised country with equally strong cultural roots, submission to authority and total and unquestionable loyalty to the one in power is time-honoured tradition. Who can ever forget the Japanese Kamikazes of the Second World War who submitted to the will of those in power and sacrificed their lives as a service to their homeland.

In Japan, the family physician is called the Shoji, who is a person of power; hence, his authority must be respected. Since most family physicians know their patients and their families almost well as they know their own, their decisions are duly respected in most matters of medical care. Submission to authority is the motto in the Japanese society and doctors have therefore enjoyed enormous respect and honour in the tradition-orientated Japanese culture.

The Japanese model may not appeal to many Western ethicists, but such is the importance of culture that one may not condemn the Japanese practice without understanding its cultural values.

In cases of rather unpleasant diagnosis such as cancer, the Japanese family physicians take the immediate family into confidence and break the bad news to them carefully and after much deliberations and detailed discussion of the pros and cons of the management, but often enough are requested by the family to tell the patient as little as possible. Apparently it is not only a mark or respect for the terminally ill patient, but also saves him from undue agony.

The major difference between the Western and the Asian models in the field of autonomy with respect to informed consent is that the close bond of the family seems to be a deciding factor in favour of the practice of shared love and decision making in Asia as opposed to the West, where everyone is unto himself. Surrogate consent is a norm in many Asian and African cultures. Perhaps in many ways it is also a reflection upon the total breakdown of the family units in UK and most of Europe.

Erosion of authority at the top invariably leads to demolition of the structure and the very fabric of the unit in question, be it a family or an institution. Besides, despite the constant display of love you sign in the West, there is so much acrimony and untold hidden secrets that the basic rules of any lasting relationship; i.e., trust, is seldom seen. One must not generalize the rule, there are some excellent examples of mutual love and family bonds, but by and large the family units of the Victorian era are a thing of the past.

Perhaps it is the tribal character of most Asian societies that may be the factor responsible for sharing everything. The fear of unknown, an invader, a warring tribe, insecurity, lack of facilities and resources, or simple fright from untoward happenings may have affected our psyche, partly explaining the bond of families in joys as well as grief.

The concept of shared feelings is nearly lacking in the West. In fact, it is considered to be almost conduct unbecoming of an authoritative person to shed a tear in public. How excessively was the Royal family criticised for not showing their emotion of grief and not sharing in with the public at large at the time of the tragic death of Diana. An Englishman is known for his stiff upper lip throughout the world, though with the decline of the empire, even that has changed much.

In the ancient days, when leprosy was considered to be contagious and plague still raged, faith unto the healers and carers and the community spirit saved many an epidemic in the West as indeed in the East. But those days are gone. Today the concept of sharing the misery is not often seen in the West, though it continues to remain a strong binding force in most Eastern cultures as we witnessed during the Tsunami disaster of 2005 and recent earthquakes in Pakistan.

Agro-based economies also seem to have strong family bonds. A farm used to be common source of economic welfare and sustenance of the family or the community. It needed hands at tilling, watering, and harvesting. These activates were group activities and a major source of development of love, affection and lasting bonds.

Of course without doubt there were some unhappy moments also when a greedy uncle would try to rob a nephew of his share or mighty landlord attempting to ransack the peasant's lot; but they were always condemned and were in fact another reason for the development of community spirit against such tyranny. Decisions were based upon mutual agreements after long and sometimes hot debates. But they were often enough a good source of inculcating the spirit of sharing and caring. Large families, herds of cattle, harvest of barley and wheat, and the joys of life all made life simple and full of happiness. Therefore, when someone in the community fell sick or succumbed to an illness, it was major incident and everyone had to join in, for sharing the grief and sorrow.

Many decisions about the day-to-day as well as long-term policies were made in the *Bethak, Autauq, or Pindal*. The seniors sat on the *charpoys* and juniors either on the ground or settees. All listened attentively to the wise men of the community. Heated debates sometimes deferred decision making, but dialogues did not stop, and finally decisions were made and conveyed to the authorities through the *Chauadhri* or the wadera. Not all of them were good, though, but despite the odd rascal, nearly all cared about their followers or else they would become as vulnerable as the one who has no tribe, clan or following.

This philosophy of shared economy and shared decision making is still in vogue in rural Pakistan, India, Bangladesh and elsewhere. The West has lost it all.

With the introduction of the Industrial Revolution, which some say began at Iron Bridge in Shropshire, the agro-based economy was gradually replaced by a fast-moving industrial economy. The race for gold led to the fast disappearance of families, communities and the societies. The rural areas were soon depleted of manpower and the menfolk left their women to fend for themselves.

The process of the erosion of authority had thus begun, which would soon be followed by the women's emancipation movement, and further breakdown of the families into solitary individuals. The concept of shared decision making was now replaced with "It's my life and I'll make the decision best suited to me." Well, the age of the true autonomy had arrived; some like it others don't, but that is how the situation exist today in the West. Not so in the Asian societies.

In most Asian societies strong family units, not just the nuclear families, take the decisions in most matters including those of health and sickness. There is usually a patriarch or matriarch, who being the wise one and enjoying total confidence, is often enough entrusted with the difficult task of decision making. It is particularly a challenging decision when it comes to the artificial prolongation of life or with-holding treatment in such hopeless situations. So it is not always easy to be the head of a family, clan or tribe after all. How true is the saying that uneasy rests the head that wears the crown.

The ancient civilisation of Iran is another example where foreign influence was accepted with some resistance and many traditions had to face modifications. Arabs always considered outsiders less civilised, hence the word *Ajami* for non-Arabs, i.e., foreigners. But Iran had a mighty, old civilisation of its own which had strong and deep roots far older than the Islamic roots in Arabia. So when Iran was finally vanquished in the days of the third caliph, it accepted the defeat, but continued to practice its own values. Iran even today has its very strong beliefs and thoughts. For instance, in the Muslim world stretching from Canada to Indonesia, Iran is the only country that has accepted, indeed recommended, cloning for research. And their modus operandi regarding abortion is also different from many Islamic countries. Believing that ensoulment takes place at 120 days after conception, abortion in Iran is allowed until just before that date. Iran, we must understand, has a very strong theological background. Some of its universities are nearly a thousand or more years old.

The religious universities in Qum are even older, and the finest in the Islamic world. Religious scholars called *Mujtahids* have to sit at the feet of the learned Ayatollahs for decades to achieve those qualifications. They are taught many sub-jects including, of course, philosophy, logic and rhetoric. They are highly cultivated and spiritual in their life, with enormous knowledge and religious experience. *Ijtehad*, i.e., research, is their main task and all matters of *fiqh*, i.e., jurisprudence, which poses many forms of challenge to ordinary folks, is referred to them for guid-ance in the light of Quran and Sunnah. Besides, they take practical illustrations for guiding principles from the lives of the *Ahal al Beit*, the 12 imams of *Ithna Ashari Shias*. In fact, Imam Raza AS, the eighth pious Imam, is buried in Mashed, which is a constant source of learning and spiritual awakening to all and sundry. Therefore, no ethical issue can receive a stamp of approval in the country unless it has been thoroughly scrutinised by the learned Ulema. Iran has continued to progress in the field of Bioethics through its perpetual endeavour to establish links with other uni-versities in the world. They also hold regular international meetings for exchange of thoughts and solving the dilemma facing the Muslim scientists on the globe.

Many things are common between Iran and Pakistan in matters of Autonomy, Beneficence and Justice. For instance just like Iran, Pakistan also has a very strong religious bond and the teachings of the Quran and Sunnah are strictly observed. A woman has more liberty in these two cultures in many ways than in the West. A fine example may be quoted in the shape of the prenuptial contact that Islam has ordained for all marriages to be solemnised under the Islamic *fiqh*. A girl, in fact, proposes the marriage, following in the footsteps of Hazrat Khatidjatul Kubra, who proposed to the Prophet through an intermediary. Likewise the prenuptial contact, called *meher,* is established before the *Nikah* ceremony to guarantee the economic welfare of the girl. The Western countries are now beginning to draw such contracts nowadays as the suing business by the estranged wives of the celebrities gets them into hot water.

Despite the freedom of a woman to choose her life partner and her career, Pakistan and many other regional countries are deeply influenced by cultural taboos. The concept of *izzat* is a time-honoured phenomenon, which basically means that a woman has a singular place in the household and she must not do anything unbecoming, which may bring bad name to the family or the tribe. It has both good and bad implications. If exploited it can mean that the head of the family may be ruthless and not allow the girl to study, let alone chose her own husband. But that is against the Islamic teaching; it is purely a cultural factor, which is based upon lofty ideas of the clans' superiority in wealth, personnel or possessions, etc. Unfortunately that is a common observation which has hindered the development of our women folk in many ways.

Translated into the terms of medical ethics, it means that the decision to undergo a procedure, treatment, an operation or participation in research may not be entirely the girl's choice, as the head of the family or at least the mother may make the final decision. But the same applies to the male members also. A young man may not choose for himself the treatment he may desire without approval of the father. It is mainly due to two factors, namely the old-fashioned concept that the seniors know it best and secondly the concept of respect for the elderly. Betrayal of trust and disrespect are two forces that often rest the decision-making power with the head of the family. Hence it is day-to-day observation that when asked, a young person would automatically reply, "My father will decide," and that means that the Western concept of autonomy must be modified to suit the cultural needs of our people.

In the villages the situation is at the extreme. Since *purdah* is observed by most women in the tribal cultures and many villages, it is nearly impossible to reach out to women to carry out any form of data collection or research. In fact, even with the help of female workers, it is often met with resistance by the male members of the society as a sort of interference in their privacy and a challenge to their authority. All of those who have practiced in such an environment would know that the matter has to be handled very carefully and delicately. Talking to the male member of the family, if tribe then obviously the tribal chief, is the first step in any form of research or treatment.

There are many exceptions, of course. For instance in the educated people living in major cities the situation is quite different though not entirely free of the basic rule of authority resting with the head of the family. However in the cities, since the young people have access to media and modern development some concept of individual autonomy has come up. Many such people would like to know more about a treatment or intervention before consenting. They would also like to know if there are any alternatives and if research is being conducted how it would benefit them or their family and countrymen.

Authority is often abused by many people who may be holding a post controlling the destiny of the ordinary folks. False propaganda to promote their cause obviously for a monetary gain may even lead them to falsify the plan, and later on the data, to justify their objectives. Many academicians have been known to have benefited in more than one way through such malpractices.

The learned philosophers have advised us that the fundamental truths are universal and moral values are supposed to be reciprocal in the East as well as the West. And yet morality is facing a major challenge globally. It is purely on account of greed, avarice and self interest that the world is going through so many crises in current times. One such example is that of the financial meltdown in USA and Europe. The bankers who were considered to be '*ameen*' or caretakers of people's hard-earned money became viciously greedy and started gambling with other people's money. That money, which was placed in their custody for safe keeping or in many cases to earn a profit to live off it as in the case of the pensioners and widows, etc., was plundered by greedy traders and stockbrokers and lost forever. The fundamental rules of banking were 'honesty and trust.' Both were blatantly exploited. In other words, the banking ethics were destroyed. Since the rules of ethics determined the depositors' money must be available on demand, when the faith in banking sector was lost, we saw the 'run' on the banks for money withdrawal at the Northern Rock and elsewhere, collapsing the bank until the government had to step in and save it from going bankrupt.

A couple of decades ago we saw one major bank in the Middle East also go turtle. Its name was BCCI and had seen some glorious days. But sad to say, that fine institution was lost because of the lack of respect and failure to observe the code of ethics by some greedy, unscrupulous and unjust people.

In the case of surrogate consent, a common practice in the Eastern culture, the same formula of trust and faith duly applies to the head of the tribe or the family also. If the head is unscrupulous or unethical, he may collaborate with some stakeholders in employing his wards in a research project or a clinical trial without the actual informed consent that each person must give.

Cultural anthropology also teaches us that individual as well as communal factors may also affect the ethical practices. For instance, in the Eastern countries altruism is seen beyond all measures, which is seldom if at all seen in the West. But the same mother who would die for her child in a poor African or Asian country may indeed be forced to sell her baby to fill her and her other family member's stomach. Exploitation of the needy ones is a norm in the world; Islam has strongly condemned all such atrocities, warning its followers time and again to refrain from harming the widows or the orphans or the downtrodden. In fact, Quran mandates clearly that the wardens may not mix up their earnings or savings with those of the wards, the orphans or the dependant folks, lest the dependants be deprived of their due share by a greedy uncle.

In fact, surrogate consent is not in the least a desirable thing, but cultural norms must be respected and since the Eastern culture believes strongly in the clannish or tribal culture, where paternalism is a norm. Therefore we must honour that practice, and keep any forms of doubts at bay. Having said this, one must also remind oneself that the Eastern family politics is riddled with stories of personal envies and jealousies, etc. One need not remind anyone of the stories of the Banu Ummaya inflicting huge atrocities on the Aha al beit and the Banu Abbas, killing their next of kin, and the later day rulers of India, namely Mughals, following suit.

4.3   Universalism or Relativism

4.3.1   An Ethical Debate

Medical ethics is going through a phase of rapid evolution. Each day brings out new dilemmas at the forefront of life and poses challenges to the experts. Since modern medical ethics is a relatively new entrant in the field of sciences, the pundits continue to explore new avenues to solve equally new and indeed some time-tested dilemmas facing medical practitioners of modern day.

In Britain, just as in many other countries, medical ethics is now being taught at the undergraduate level. Since these are early days, much needs to be done before a definite syllabus can be finalized, though it is doubtful if it shall ever be. After all, medical ethics is new, highly innovative and dynamic. Each day brings in new challenges, and medical graduates can only be taught so much, and the rest they have to learn through direct exposure to the problems. A modern medical graduate must be a lifetime learner.

One such debate seems to focus on the philosophy of Universalism versus Relativism.

The principle of Universalism preaches the practical application of universal code of medical ethics to all nations, regions, countries and continents, indeed the entire globe.

The principle of Relativism, on the other hand, advises us that although the basic principles of medical ethics are universal, they must conform and should be modified to match the local customs, traditions, culture, and religious faith of the population in question.

Social anthropologists inform us that the impact of social elements, particularly culture and faith, is a major factor in determining the way of life of an individual as indeed the community. Culture is like a cloak covering the body of a person or community and its inner core, which comprises faith or beliefs. Like in the past, there are a myriad of cultures that do not have a particular faith or religion, but they do not form the major bulk of the population mass; and even they in one form or another have some form of belief or faith. The pagan festival of the Stonehenge in England is a fine example of the concept of ancient religions going several millennia back into history seeming to fulfil the believers' innate desire to believe in something. They have their own culture, which is deeply immersed in their customs, traditions faith and beliefs. E.B. Taylor, a famous social anthropologist, in 1871 defined culture as "The complex whole which includes knowledge, belief, art, morals, law, custom, and other capabilities, and habits acquired by man as member of society."

Universal principles of modern ethics were not established by modern philosophers, but by the ancient Greek masters like Plato and Aristotle. Some of these mighty thinkers had some form of religious faith, others did not. And yet the principles laid down by them are universal, Truth shall always remain a virtue; so would honesty, bravery, courage, generosity, and of course knowledge. Religion or not, the virtue cannot be belittled by a nonbeliever, an agnostic, or an atheist. Only an evil person may attempt to undignify virtue and such a person may infact be a believer.

The wise men of ancient China, and the savants of ancient lands of Indus, i.e., India, and the lands of Nile, and similar ancient civilisations of Mesopotamia and Babylon of Hammurabi and Nebuchadnezzar, or the Incas, all had their codes of conduct, which they followed in their day-to-day life. To most cultures the dignity and honour of mankind, respect for the rights of an individual and similar attributes desirable in a person or a community were holy.

Although it is true that one does not have to be a believer to be an ethical person, it is also true that the divine commands have streamlined the principles of ethics in all those societies that have been fortunate enough to receive them. It is almost like saying that you may cross the road anywhere, but if you do so at a prescribed site such as a zebra crossing, the chances of meeting an accident would be smaller.

Religions have undoubtedly clarified the grey zones in the matters of good and evil and have continued to guide us over several centuries. All religions have common principles of human dignity, autonomy, human rights, duties, and obligations to an individual and the society, etc. Both thought as well as action are guided by the religious code of conduct.

Islam is the final religion to have brought the last message of God Almighty though His last prophet. Therefore, its message is universal, lasting and global. The message of Islam in matters of ethics can be summarized in the simple but powerful terms, namely *Al-Takreem, wal Birr, wal Adl, wal Ehsaan*, i.e., autonomy, justice and beneficence. The latter-day development of the principles of modern medical ethics by some Western Ethicists into the famous four principles, namely Autonomy, Beneficence, Nonmaleficence and Justice, have indeed surfaced at least 1,400 years later than they were revealed in the holy Quran.

In the context of present debate, allow me to say that Universalism is often described as 'ethical imperialism.' It may not be far from truth, if one examines the impact of universalism minutely. What the portents of Universalism say is that the rules of ethics applicable to the Western societies must equivocally apply to the Eastern societies also. It is almost like dictating the rules developed for the West by the Western philosophers to apply in culture which has far deeper insights into matters of ethics than any modern-day Western culture or its philosophers. Hegemony has different colours and many shades indeed.

On the other hand, one cannot deny that many, indeed most, principles of universalism are universal, and do not require endorsement by any Western philosopher to receive approval before implementing them globally. Some examples have already been quoted like truth and justice, honesty and fair pay, etc. They are universal truths and virtue applicable across the board. However, sometimes interpretations of these fundamental principles may vary, depending on the motives of the mighty and the powerful. What may appear to be just to the victor may truly appear to be injustice to the fallen and so on.

Autonomy is the spine of medical ethics and the informed consent is akin to the spinal cord. Although paternalism has been nearly discarded in the West, perhaps it still has some place in the developing nations of the world. Education has brought the knowhow to the individuals in general and more specifically to the West. But in a society where education and literacy are still struggling to even reach any noticeable levels, the patient has to put his faith and trust in the physician. The physician in his own right must not ever exploit his position thus granted to avail of unethical dividends. Thus paternalism has a very major role to play in our societies. Likewise, since many women and children do not have same access to the outside world as the menfolk, surrogate consent also has some place in the developing nations of the world. Thus here we have a paradox with the Western concept of autonomy. Let us now examine the impact of Relativism on Informed consent in the light of social anthropology.

In the West education and economic independence are two major factors that have deeply influenced the way of life and decisions made on daily basis. Not long ago, British women were did not have the right to vote, let alone contest a seat in the parliament. They have always received inferior wages for the same jobs as compared to men and do not have top positions in the corporate sector, except a few. Daily reports of domestic violence on the part of a drunken husband or partner and an occasional report of a murder committed by an angry male member are not unheard of. Until quite recently the fabric of the British society remained intact where a husband took care of the economic matters and the wife of upbringing of the children. No more. The very fabric is tattered and torn. More and more people produce children outside of wedlock and such children are given the title of love child. Now if such a principle is applied universally on the basis of Universalism, how many societies would accept them in the East or indeed even in the West?

Emancipation and economic independence have now almost equalised the status of women to men in the West. Living together temporarily rather than in a lifelong commitment, i.e., 'till death do us part' and similar stories are all but folktales or at least threatening to be. The secure environment of a home so essential for a child's growth and development is lacking so badly that every day we hear of the tragic tales of a battered child or a child left to the mercy of the caretakers.

In most developing countries in Asia, Africa, and Latin America, despite hunger and disease, the sanctity of the family unit is still maintained, and children are protected beyond personal safety. Some fine illustrations of altruism are seen in these countries in which a mother would do any thing to save her child and a father even sell his kidney to buy food for his young ones, a practice not to be condoned but nevertheless witnessed in India and elsewhere on fairly regular basis.

The concept of a nuclear family is still new to most developing countries. Extended families are still a norm in most societies. They may have many ills of their own, but in the hour of need no one but no one, except your own kith and kin, come to shelter you, a story if ever seen in the West.

In the Eastern culture marriage is a sacred covenant solemnised in the mosque or mandir, a Gurdwara, or a church in the presence of many people and taking God as a witness. It brings dignity to the girl, a home for both man and woman, and a nest for the child to come. The family unit is thus like a barrier that a gardener erects around a tiny sapling to protect it from the harsh winter, the soaking rain or the scorching heat of the sun. Thus protected, the child grows to develop into balanced individual, useful and productive in his own right.

In some societies like Pakistan, India, Iran, Afghanistan Bangladesh and most Arab countries, women observe Purdah. It involves covering the head with a scarf and sometimes the whole body with a chador or Burqa. It is an indicator of human dignity and a symbol of modesty. Sadly enough, circumstances have changed in many countries and economic needs have forced women out of the Chador, to meet the demands on the family commitments. But given the choice many would still prefer to cover themselves when going out.

The primary task of women in most developing countries is to take care of the domestic chores, while the male members go out to work and earn the living. Authority often rests with the male members of the society, but not always and not everywhere. In Malay it is often the women who make the final decision in most matters, including health care.

In many parts of Pakistan and elsewhere in the region, the society is basically tribal in character. The tribal culture therefore determines the parameters of all daily as well as long-term matters. The Baradari system is so deeply rooted in these regions that despite education and the exposure to the media, etc., certain norms have not changed one bit from the day they were introduced several millennia ago. One such subject is the matter of honour.

It is difficult to define honour per se. It is everything to a proud Baloach, Pashtoon, Afghan, Uzbek and many others of the tribal communities. It is the matter of honour that has resulted in many wars between the tribes sometimes lasting several generations. It is also the matter of dignity and respect to the individual members and to the community at large. It also means respect of the elderly, the care and love given to them and the status accorded them by virtue of seniority. And above all honour also implies protection of women and children from any form of harm, either from within or outside the community.

One such example of sanctity of honour, in respect to the care of women, appeared in a rather negative form on the British TV in July 2007. A Kurd girl married an Iranian, who obviously was an alien and outsider. The father and his accomplices killed the girl in cold blood, and without any remorse, and went to gaol for life to avenge the family honour. In fact, honour killing is also seen in many other communities like those he Sikhs and Hindus. It is not so much the matter of religious faith as of the tribal values.

Karo Kari is a not too unfamiliar a crime often witnessed in Sindh. It has more to do with the tribal values than the religious faith and it is more likely to be a secular, almost pagan, practice than religious in any way.

Since culture is dynamic and continues to evolve with the circumstances, its impact on medical ethics also remains ever changing. And within a given culture there often are several subcultures, each having its own influence. Sometimes each subculture jostles with the other similar but somewhat different subculture, resulting in undercurrents. The sum total of all such efforts is usually summative in nature and often enough uniform in pattern. Culture is often heterogeneous and therefore results in conflicting outcomes for a person who may not be fully conversant with the given culture or cultures.

Culture can also be expressed as the concept of applying certain principles in matters of individual life and hereafter or that of a community. Many of our cultural influences may be imperceptible and non tangible, but most are rather apparent and quite distinctly visible. A study of culture is almost mandatory in medical education, because what may appear to be a genuine matter of greeting, such as a peck on the cheek of a child, may be quite unacceptable to another member of a different culture within the same community. In Malaysia, for instance, there are no call bells in Malay homes, as it is considered rude to shake up the resident of a home with the unpleasant sound of call bell. Instead it is customary to simply call out the name of the person in gradually raising level of voice three times only. If the person whose home you are visiting does not reply to your call of *Salam o Alaikum*, you are obliged to quietly go away.

The differences could be vast between the similar communities, let alone between the West and the East; Universalism would meet the global needs in matters of medical ethics. However, Relativism appears to take into account such factors as cultural differences, religious components, traditions and many other similar tangible and non-tangible variables. And it appeals to the common sense.

4.4   Medical Ethics in a Pluralistic Society

Practicing medicine in a pluralistic society can be challenging. Formal teaching of medical ethics in Britain is relatively new. The Universities are actively engaged in identifying the relevant teaching material as indeed the appropriate tools. One essential component of the GMC's vision of *Tomorrow's Doctor* has now begun to materialize.

From ancient times medical knowledge and technology has been transferred from generation to generation through long years of apprenticeship. Long-term close contact between the master and the pupil almost imperceptibly enables the young to imbibe the information from the old. It also involves learning the etiquette of dealing with the patients. Such mannerism revolved around the basic attitudes like greeting the patient, respect and empathy, etc. These attributes are expected of any Good Samaritan. The life was peaceful and the profession did not face any formidable challenges. The doctor knew it best. The paternalistic approach worked well. It was a time of mutual trust. However, with the explosion of the knowledge and advancement of technology, the physician is now exposed to an ever-growing myriad of ethical dilemmas, particularly so in a multicultural society.

The patients coming from different cultural backgrounds expect to be dealt with in conformity with their norms and practices. There are many examples of the impact of diverse culture on bioethics. For instance, an Englishman likes to know everything about the problem, plan of treatment, risks and complications, alternate therapies, etc. He alone must take an informed decision in all matters of his illness. A South East Asian patient, on the other hand, often reverts to the family patriarch.

Each culture has its own values. For instance, there is a great diversity amongst the Malaysians. Those of the Indian stock would leave all decisions to the patriarch, the Chinese take all decisions at the masculine level, and most ethnic Malays would leave all matters to the homemaker. Surrogate consent is often the standard practice in most Asian communities. The Filipinos tend to be very polite, soft-spoken and independent decision makers. No wonder the top cruise liners hire them on their luxury cruise ships!

The Afro-Caribbean communities on the other hand are very different compared to the British Asians. They are a self-assured, strong-willed and quite independent people. They almost always take matters in their own hands. The Polish and many Eastern Europeans who now constitute a formidable portion of the immigrants, have very different values, not only deeply rooted in Christian ethics, but also strongly embedded in their rich cultural heritage.

When it comes to the question of confidentiality, we are taught not to reveal the patient's findings to anyone, not even to the spouse. And yet more often than not, when an Asian elderly person, who may have a serious illness, walks into the clinic, the situation is typical of that culture. He is often preceded by a close relative, insisting on being informed *first*, pleading not to break the bad news to the patient to prevent a crisis. Perhaps many consequentionalists may like this approach.

Suffice it to say that there are lots of challenges that one may face in day-to-day's work, and there are no strict guidelines currently available to deal with multiethnic patient population. The standard guidelines published by the GMC and other authorities from time to time are invaluable in providing the essential information and advice in most matters, but are they tailor-made to suit all cultures and all patients coming from different backgrounds? Probably not.

The answer lies in laying more emphasis upon formal teaching of medical ethics in the medical school. Theoretical knowledge of normative principles taught through didactic lecture may be supplemented with real or simulated scenarios of multicultural ethical dilemmas in the skill labs. In the clinical years it could be a lot easier, particularly in most city hospitals, which have a fair share of multiethnic patient population. Clinical workshops, symposia and seminars could be other forms of tools of learning at both preclinical and clinical levels.

It has been rightly pointed out by some workers that "to contribute usefully to contemporary debates, ethicists need to better address the multiethnic, multi-faith character of contemporary social settings [1]. They need to recognise the existence of a plurality of communities, of interpretation and local moral worlds" [2].

The first generation immigrants have all but settled into retirement and form a fair share of day-to-day visitors to the GP surgery. The new generation, though thoroughly British, still maintain their ties with the former home country, through marriages, social visits, or religious activities.

It has been recently pointed out that "for many of those who trace their origins from other than a European lineage, it is clear and distressing at times that the NHS has found it difficult to adapt to the needs of minority groups" [3]. Rapid advances in knowledge and the lack of awareness of the religious and cultural values as well as their specific ways of life or rituals of death, are posing different kinds of challenges.

Regrettably many answers may not be found in the most popular text book on medical ethics called the *Principles of Biomedical Ethics*, written by Beauchamp and Childress [4] or by following the philosophy of the books or the standards of Good Clinical Practice only. They are undoubtedly invaluable in laying the foundation, but the building must be built through practical experiences. Such an exposure, in fact, should be given to the medical student quite early in his life. As a matter of fact, the decision making in many matters of life and death may require at least the basic working knowledge, if not more, of the faiths and beliefs, rituals and practices of the community that a doctor attends to.

Since the British medical care was essentially designed to cater to the needs of a homogenous population, it should not be surprising to note that some of the themes considered as norms may not be suitable in many a patient with a different background; "central to discussions concerning ethics, and medical ethics in particular, there must lay an understanding and an appreciation of the beliefs, perspectives, and conceptual frameworks used by our patients" [5].

Garter and Sheikh [6] have published a paper highlighting the principles and practices observed by the Muslims, who in fact form a sizeable component of the United Kingdom. "The 1991 Census revealed that almost 6 % of the UK population classified themselves as belonging to a minority ethnic group" [7]. The latest edition of a book called *The Life in the UK*, mandatory for anyone applying for naturalisation, quotes the figure at 9.2 % [8]. *The Daily Guardian* published a news item on April 8, 2008, quoting the former home secretary Jacqui Smith on her visit to Islamabad that the Whitehall believed that the Muslim population in Britain now stands at 3.3 %, or the two-million mark.

Garter and Sheikh [6] point out that "A minimum level of cultural awareness is a necessary prerequisite for the delivery of care that is culturally sensitive." They have highlighted some of the fundamental issues, such as genetic manipulation, assisted conception, adoption, prenatal screening, abortion, end-of-life issues, organ donations, and many other ethical questions, in the light of relevant religious beliefs and scriptures.

It appears that in the debate on these issues, the concept of 'identity' is of crucial importance. The burning issues are 'authenticity' and 'integrity'. What authenticity entails is a responsibility to live our lives in accordance with the values that constitute our unique cultural perspective. People also need to retain their integrity as a group. They must uphold the shared values that unite them [9]. In a colourful society of today's Britain, an Asian, an African and a European can all maintain their authenticity without compromising the integrity as a member of the British nation.

It is obvious that in a multicultural, multi- faith, pluralistic society, religion, customs, norms and values determine both the genuineness and the authenticity of one's perspective.

References

1. Chan JJ, Chan JE (2000) Medicine for the millennium. The challenges of the post modernism. Med J Aust 172:332–334
2. Kleinman A (1995) Writing at the margin: discourse between anthropology and medicine. University of California Press, Berkley, pp 41–67
3. Mackintosh J, Bhopal R, Unwin N, Ahmad N (1998) Step by step guide to epidemiological health assessment for minority ethnic groups. University of Newcastle, Newcastle, pp 1–22
4. Beaucahmp TL, Childress JF (1994) Principles of biomedical ethics, 4th edn. Oxford University Press, Oxford
5. The Department of Health (1999) The patient's charter. DOH, London, p 4
6. Gartrad AR (2001) Sheikh a medical ethics and Islam: principles and practice. Arch Dis Child 84:72–75
7. Anwar M (2000) Muslims in Britain: demographic and socio-economic position. In: Sheikh A, Gatrad AR (eds) Caring for Muslim patients. Radccliffe, Oxford, pp 3–16
8. Life in the UK. The official document of the British home office. Valid from 2007. p 37
9. De Castro LD (1999) Is there an Asian bioethics? Bioethics 13(3/4):23, Blackwell Publishers Ltd

Ethics and Evolution of Medicine

5

5.1 Research Ethics

Research is the backbone of every form of knowledge, be it religious or worldly. *Ijtehad* is the fine line that differentiates *Imamia* fiqh from other *Fiqhas*. *Ijtehad* is what all those learned scholars, Ayataullahs and *Maraajae*, do throughout their lives in such eminent places of learning as the *Hawza-e Ilmia* of Najaf Al *Ashraf* and *Qum*. The final product that emerges from these institutions are called *Mujtahids*, experts and super-specialists in the *Ilme Din, Fiqh, Ilme Ahadith,* and *Ilmul Kalam, Ilmul Rijaal* and many other sub- and super-specialties. Their task continues throughout their lives as they guide Muslims in all matters of Din and complex issues of Sharia throughout their lives. Some, like Imam Khomeini, believed in *Wilayate Faqih* and therefore, besides religious matters, they guide Muslims in political matters too. Others, like *Ayataullah- Ul -Uzma* Syed Sistani, prefer to remain isolated from politics. But they are sources of guidance in all religious matters.

Research ethics is a special subject. Islam grants undue unimportance to seeking and furthering knowledge. The famous *Ahadith* in this respect needs no further reminder. Suffice it to say that not just the Prophet and his *Ahlebeit* but also the Quran clearly mandate that every believer pray 'Rabbe Zidne *Ilma*'. Instead of asking for ransom money, the Prophet ordered the educated among the prisoners of *Badr* to impart education to the Muslims in order to regain their freedom. That is one practical example of the importance the Prophet accorded to education.

In the field of medical sciences, extensive research goes on throughout the year and across the globe. Since research demands massive funds, fundamental research is often carried out in rich countries, among which America stands at the top, followed by other Western countries and now some developing nations like China and India.

A consultation was organised with leading medical practitioners in Geneva in October 1994. The subject was the teaching of medical ethics. The participants in the group agreed with the notion that teaching medical ethics is essential to train young minds to learn ethical principles and their application in their professional

S.H. Zaidi, *Ethics in Medicine*,
DOI 10.1007/978-3-319-01044-1_5, © Springer International Publishing Switzerland 2014

life at an early stage of their career. It also was agreed that medical ethics should be integrated into the undergraduate curriculum of medical schools around the globe. The group also reached a consensus that medical ethics should be taught through interactive seminars, Problem Based Learning (PBL) and data-based studies and evidence-based medicine.

No doubt, medical school is the first nursery for a medic to learn the fundamental principles of medical ethics. Medicine is changing every day; new discoveries are made, and by the time a book is published, it is already outdated. Research is an integral part of medical education. No new invention or discovery is possible unless a hypothesis is developed, followed by a detailed scientific study in a laboratory, followed by study using animal models, then voluntary human subjects and finally patients, under absolutely controlled conditions. Single- and double-blind studies are conducted by research scientists and physicians under strict controls. Phase 1 and 2 trials are conducted in pioneering labs by a group of highly dedicated scientists with doctorates in their chosen specialities. These phases may take years, during which a scientist may spend hours of lonely existence with his or her mice or guinea pigs or just mathematical calculations trying to reach a definite plan of action or prove the hypothesis. Once a hypothesis is proven it changes into a thesis; if it fails it becomes an antithesis. Sometimes disproving a null hypothesis may indeed be more rewarding than proving a hypothesis.

Once the lab and animal experimentation is complete and acclaimed by the pundits in the field, often published in an indexed journal with as high an impact factor as possible, phase 3 and phase 4 trials would begin. Many such trials may be carried out in the country of origin of the molecule or the theorem, but they often are carried out in developing countries, where demand for many drugs is highest, particularly antibiotics. In addition, the reach of trials conducted among the targeted population would be better accepted by medical professionals than results from an alien country with different biological and anthropomorphic characters.

Jabir Ibne Hayan, the father of alchemy, is said to have narrated that he was encouraged by his teacher Imam Jafar -al- Sadiq to conduct experiments and do research in the field of alchemy and the Imam would then check his results. Many times the Imam would advise him to go back to the lab until a result was achieved of which the Imam approved. Jabir called the Imam 'a mine of wisdom; and his teacher.'

Knowledge without research or its propagation becomes stale and quickly outdated. *Ijtehad* is clearly mandated in *Fiqh Jafari* as the famous *Hawzae Ilmias* of Najaf and Qum have continued to enlighten us for a thousand years.

Islamic history is inundated with glittering stars like Al Kindi, Al Farabi, Bu Ali Senna, Al Tabari Al Khayyam and modern scientists like Nobel laureate Professor Abdul Salam, who have all promoted research and propagation of knowledge and encouraged others to follow.

Animal experimentation is a standard practice, as a new molecule or therapy needs to be used on live models before employing them on human subjects and finally patients. Without animal models it may not be a possible to establish whether a particular molecule developed by scientists in their rather reclusive atmosphere is useful to mankind.

Many pro-life activists and animal rights protectionists in the UK and elsewhere raise their voices from time to time against animal experimentation. In fact, a few years ago they gathered near Oxford outside an animal farm that was supplying guinea pigs and rodents to the famous Oxford labs; they destroyed some of these farms, causing much damage.

The scientists who carry out such experiments apply the ethical theory of utilitarianism to defend their actions. By using animals, in their view, they are saving mankind from experimentation, as indeed with the clear motives of bringing relief from pain and suffering caused by diseases.

Research involving human beings did not stop after the infamous Nazi experimentations in prison camps. It did, however, bring into focus the atrocities conducted by evil German doctors in Auschwitz and Dachau, under the Utilitarian theory, that is, to harm a few to save a million. That was nothing more than falsifying the fact. The truth of their atrocities was revealed at the Nuremberg Trials, when it was disclosed that Dr. Mangle et al. immersed many subjects in boiling water or immersed them naked in freezing cold water to study the effect of extreme temperatures on the human body and measure the degree of human tolerance to pain and torture. They also castrated many young male prisoners and performed oophorectomies on young females to study the effects on their physiology. All these experiments were performed without any form of consent, through force, coercion and utmost disrespect for human rights. The torture these prisoners suffered at the hands of Nazis makes one see why Ali As once said, "There are times when death could indeed be the sweetest of all things for a person."

The Nuremberg Trials resulted in the promulgation of the Nuremberg Code, which was approved in 1947. The founding principle of this code was that in all human experimentations 'the voluntary consent of the human subjects was absolutely and unequivocally essential.'

Since the words in the Nuremberg Code were those of the lawmakers rather than the physicians, many pitfalls were found by later generations. One such fault was the absence of specification of such human subjects who may not be in a position to give informed consent, such as the mentally disabled, drug addicts, and those who were destitute, dementic, intoxicated, the minors or comatose. In 1964, therefore, the World Medical Association (WMA) held a meeting in Helsinki, Finland, and developed the 'guidelines for ethical human research.' It had 12 guiding rules to follow.

Chronologically speaking, following the Nuremberg Code, the Council for the International Organisation of Medical Sciences (CIOMS) and the WHO adopted a declaration on the subject of human experimentation. This code of ethics was then adopted by the 9th World Medical Assembly in Helsinki in 1964. It was called the 'Helsinki Declaration.' This declaration is periodically updated. It was updated in 1975, by the 29th World Medical Assembly in Tokyo. It was called the Helsinki Declaration-2. This document has been updated on several occasions since. The latest version is accessible on the internet: Declaration of Helsinki—World Medical Association, available at www.wma.net/en/20activities/10ethics/10helsinki/Helsinki and the Declaration of Helsinki. *World Med J* 2008. 6th rev.

5.2   Human Experimentation

Many Western scientists collaborate with their counterparts in developing countries in conducting useful and extremely beneficial clinical and medical research. Some of them involve human beings, who are thus aptly described by some as human guinea pigs, a rather derogatory term, nonetheless often enough used by many. Most of such experiments and trials are quite ethical and above board, but some are highly controversial and unethical, as we shall see in the following essay.

For many years, Thailand and the Philippines have been the target areas for many human trials, though Africa is not far behind. Lately, however, more and more research is being carried out in China and India. The countries in Latin America as well as the South Pacific islands are also in the news.

Undoubtedly most of the work carried out on human beings must have been approved by the competent health authorities of the relevant countries, before actually seeing the light of the day. But unfortunately, many developing nations do not have extremely bright reputations in terms of honesty, integrity and fair practices. Lately Nigeria is rated as one of the most corrupt countries, though unfortunately enough Pakistan is not far behind in the latest survey published on the most corrupt countries in the world.

Insecurity and lack of justice are two major reasons for the greed in these countries. It is often described as a third world phenomenon. Even in Karachi, one may have noticed that the driver does not bring the gear back into neutral position, let alone apply the hand break at the red light. And his car is always moving and pitching to and fro. Why, because of the lack of security and uncertainty, lest the signal may fail, lest the traffic cop may target him for a quick buck, or simply the haste to overtake. The concept of forming a line or the queue is unknown in most of our countries, as yet again one does not know if waiting in the queue may deprive him his share of good luck in obtaining his share of said ration or a goody, etc. Hence it is an established phenomenon in our nations that due to uncertainties and insecurity, one is in a state of constant hurry and race to reap the richest rewards as fast as time would allow.

In contrast, Western societies have established norms. No one is in hurry as he knows that he will get his due share at the appointed time and he will not be deprived of his right. Justice is more often than not granted to each and every one irrespective of the differing variables.

Not always, though. There are many examples of gross injustice, inhumanity and outright atrocities that the West has carried out on human beings and continues to do so. Let us not forget the recent happenings in Iraq and Afghanistan. But they are political matters and that is not our subject; nonetheless, those perpetual atrocities are mentioned here as the means of constant condemnation of those responsible for them.

Similarly, one must condemn those scientists who carried out the human experimentations without informed consent of the victims in Auschwitz, Treblinka, and elsewhere in the Nazi era of the pre-second Great War times. But that was in relatively old times when tyrants and ruthless dictators ruled the world, or at least endeavoured to rule it.

How about in the year 2003, when the British news media disclosed that the scientist at a renowned hospital, were discovered to carry out stealthy neurophysiologic research on human brains removed at autopsies without informed and written consent of the victims or their next of kin? Theft of a human brain remained quite a story for some time until it disappeared in the dark alleys of time. So we know that Einstein was not alone when he lost his brain to an unscrupulous scientist; there were many other mortals who met the same fate.

Liverpool is famous for the Beatles, football and very famous university hospitals. One of them is the Alder Hay Children's Hospital. In 2001–2002, it was suddenly disclosed that a large number of organs were removed from the dead bodies of children, over a period of three decades by a professor of Pathology, for research and experimentation, without the consent of the parents or the guardians of the dead babies. Once exposed, the scientist in question quietly disappeared from the scene back to his home country in Europe, and since then we have not heard much about the episode.

Throughout the world, scientists are engaged in carrying out human experiments, for without such research, science cannot advance. But such research or any kind of work involving human beings must be cleared by the relevant ethics committees and must be acutely monitored by a body of experts who must not have any self-interest in such matters.

Nevertheless, human experimentation goes on without the necessary aforementioned requisites.

In fact, while the Nazis were perhaps still contemplating human experimentation, the Americans had already launched their infamous Tuskegee trials that shall bring them eternal shame and force the 42nd President to publicly apologise to the remaining victims of Alabama trial.

In 1932, the public health services of the USA launched the Tuskegee Syphilis Study to investigate the outcome of untreated syphilis. This study has been called one of the darkest spots on the human conscience. It involved about 400 plus African-Americans living in Macon County, Alabama, suffering from syphilis. Most of them were ordinary folks, ignorant, illiterate and uneducated, nonetheless human beings and free citizens of United Sates of America. All these people involved in the study were given to belief that they were receiving the finest and the latest treatment available, while they were only being given the placebos.

In 1940, penicillin had been discovered and was a certain cure for syphilis. However, no penicillin was given to this group, so as to study the long term effects of untreated syphilis. At least 300 of these people succumbed to syphilis, suffering the most miserable long-term effects that are known to any physician. Penicillin, though available, was deliberately withheld from these unfortunate victims, and they continued to suffer the miseries caused by syphilis. To add insult to injury, a sum of $50 was given to the family in the name of funeral expenses, in lieu of autopsies performed.

This experiment lasted for 40 years, far after the illustrious Nuremberg Trials of Mengele and the lot and many years after the Nuremberg Code had been in practice. The whole world knew of the atrocities of the Polish camps, thanks to the powerful Jewish media, but no one knew of the plight of the blacks in Alabama until 1972.

Peter Buxman, a public health worker in that year, could no longer take the burden on his conscience anymore and exposed the ongoing experimentation. The hungry media reaped rich harvest and jumped onto the bandwagon, thus letting the world know of what went on from 1932 until 1972, on the black Americans in Alabama. The study was obviously terminated and the usual apologies and the drama of investigation into who did it went on for awhile. President Clinton publicly apologised on the television to the African Americans who were exploited, mistreated and abandoned to suffer in the name of research.

That was in modern-day America.

Undoubtedly, advancements in medicine depends upon scientific research, which is initially hypothetical and theoretical, then experimental and finally applied. Research is mandatory for advancement of knowledge and skills and for the welfare of mankind and society. Many studies remain only theorems, some become observational studies, some require data analysis, some involve extensive literature search, experimentation analysis and so forth. But any study that requires involvement of human beings has to be very meticulously designed, developed and carried out. It must involve the ethical approval of competent authority and never without a fully informed consent of the person or the party involved.

But that is exactly what did not happen in the case of Baby Fae. In 1984, Baby Fae (not the real name, out of respect to the family) was born with congenital hypoplastic left heart syndrome in California. She was seen in the famous Loma Linda University medical centre, by a team of Paediatric cardiologists. After much deliberation, observation and planning, Baby Fae was taken to the operating room, where under the able supervision of a famous cardiologist, a heart transplant was performed. But it was not a human heart that was transplanted, but that of a baboon. It was the first-ever experiment of a xenograft in the heart in a baby. Prior to this experiment many other attempts had already failed. Following the xenograft, Baby Fae survived only 20 days.

Later on, several questions were raised by the experts, some even daring to challenge the very concept of experimental surgery on minors. It was also questioned whether or not the mother was given adequate, wholesome and unbiased information about the procedure, because she was an under-privileged, single parent belonging to a minority ethnic group.

One of the cardiologists on the team answered some of the questions raised, even justifying the experiment on the basis of wealth of experience he possessed. He said that he had performed more than 150 organ transplants on animals, and that a human heart was not available at the time for Baby Fae, clearly admitting 'an oversight on our part not to search for human donor from the start'. As to whether or not the same formula could have been applied in the case of a similar baby of a more privileged background remains unanswered.

Following the infamous Gulf War of the1990s, many British veterans complained of unfamiliar and bizarre symptoms, some even blaming the immunisation that they had to receive prior to the trip to the Gulf region 'to protect them from a possible biological and chemical agent' allegedly developed by the Iraqi army. Such experimental vaccination, obviously, was carried out without fully informed consent or such questions likely would not have been raised afterwards.

The story of organ theft at the Alder Hay Hospital, Liverpool, has continued to echo in the corridors of health authorities for some time now. As a consequence of the discovery of that heinous practice, it is now mandatory that all tissues removed from the human beings at the time of surgery must be clearly labelled as 'for histology,', 'to be destroyed,' etc., if no further investigation is required. Retention of any human tissue is not allowed at present. If one has to save an organ or tissue, informed written consent of the patient or the ward is mandatory.

Despite all these measures, surely there would still be some scientist somewhere lurking in the dark shadows of a lab, or a corridor, contemplating some sinister activity, more for his self-interest than for the general welfare of mankind. After all, man is egoistic and much research is carried out, not so much for the fundamental reason for research but for personal glory and self-projection. Who knows where to draw a line between the good and the evil; if you are the judge, jury and the hangman all in one. Therefore, it is now compulsory that all research proposals must be judged and scrutinised by independent authorities and ethics committees before a grant is sanctioned or a research is allowed to be carried out.

It is the exercise of constant and regular reminding of the unhappy episodes of the research related to human beings that has kept many unscrupulous persons at bay; otherwise, many more Tuskegees might happen, and who knows if they are actually happening at this very minute?

Here is an illustration of abuse of authority and unethical use of placebo in drug trials. The full report is documented on the internet and is self explanatory. (http://actgnetwork.org:1996; revision.)

'The AIDS Clinical Trials Group (ACTG) Study 076 of Zidovudine in maternal-infant transmission of HIV had been published in 1994 (Connor et al. 1994). This was a placebo-controlled trial which showed a reduction of nearly 70 % in the risk of transmission, and Zidovudine became a de facto standard of care. The subsequent initiation of further placebo-controlled trials carried out in developing countries and funded by the United States Centers for Disease Control or National Institutes of Health raised considerable concern when it was learned that patients in trials in the United States had essentially unrestricted access to the drug, while those in developing countries did not. Justification was provided by a 1994 WHO group in Geneva, which concluded "Placebo-controlled trials offer the best option for a rapid and scientifically valid assessment of alternative antiretroviral drug regimens to prevent transmission of HIV" [2]. These trials appeared to be in *direct conflict with recently published guidelines (Levine 1993) for international research by CIOMS, which stated "The ethical standards applied should be no less exacting than they would be in the case of research carried out in country"*, referring to the sponsoring or initiating country [3]. In fact, a schism between ethical universalism [4] and ethical pluralism [5] was already apparent before the 1993 revision of the CIOMS guidelines (Levine 1993)'.

Last summer, an adult congenital cardiologist from my family in the United States showed me a video which shook up my belief in human justice in the most advanced nation. It is called 'How the Lord made.' It is a sad story of an African-American lab technician in an Ivy League university hospital, who despite doing groundbreaking work in developing a shunt for treatments of blue babies in animal models, was deprived of any recognition, even though his technique has since saved hundreds of lives. The man was not even allowed to enter the hospital from the main entrance. The whole credit was taken up by a famous American cardiac surgeon, and the one who actually put the idea into practice was discarded into wilderness for many long years. The story is of late 1950s, and one is simply flabbergasted at the merciless injustice and bias that goes on even in the most renowned medical institution. It is a typical example of the abuse of authority.

Now let me quote two examples of abuse of knowledge here. One in the most developed country, namely Britain, the other in the most underdeveloped territories bordering Afghanistan and Pakistan.

A few years ago, a renowned child specialist working in the NHS published a series of studies. They were mind- boggling indeed. He claimed that MMR, the conventional vaccine that the world had been using since the second World War, was leading to autism. So much scare followed these papers that the GMC had to step in. A whole generation of children were deprived of the vaccination as the parents feared them developing autism. Eventually the doctor was found guilty of fraudulent studies and sacked from NHS and deregistered from GMC. In the spring of 2013, a small epidemic of measles broke out in Wales and elsewhere, as these children had been deprived of the preventive vaccine as a fallout of wrong reporting by a famous doctor. The massive campaign to vaccinate young teenagers was launched to avoid further spread and save the children from innumerable complications of measles.

So here is an example of a negative, misguided or outright dishonest research leading to a plethora of problems in a highly developed society.

Many years ago, when WHO started a campaign to overcome the pandemic goitre seen in the Himalayan territories of Pakistan, the mullah didn't like it. He went on the hunt in every village, in every mosque, to condemn the addition of iodine to their table salt. The mullah, being the worst enemy of mankind let alone religion, spread all kinds of rumours including the one that was really effective in dissuading male folks to consume iodised salt. The mullah led them to believe that the Iodised salt provided by the 'Feringhe,' i.e., the foreigner, would lead to sterilisation, even impotence.

The worst example of mullah misguiding people was that the Feringhe has been after the mountainous lot of Afghanistan, and present day Pakistani borders since the eighteenth century, so much so that in connivance with the government they have built huge hydropower dams on the Indus such as Mangla and Tarbella. The mullah would hit the nail on its head by claiming that it was not just the polluted salt they were feeding you but destroying your religion also, as the Wudu with the water coming out of the dams was haram. Because they had taken out its '*bijli*', (the electric power), so the water was *dead*, hence *haram* for ablution.

So ignorant and gullible are the masses in those regions and so abused is the religion that one is not in the least surprised to note the atrocities that mullah has brought on in all these countries through brainwashing of young minds, cultivating them into suicide bombers destroying world peace.

The tragedy of polio vaccination teams in 2013, facing murder, rape and outright denial of access to children in the northern territories of Pakistan is much too fresh to even comment upon. Only three countries are now infected with polio virus. One of them is Pakistan, and all due to ignorance and misguidance of the masses by the local mullahs.

One is bound to bow to the wisdom of Plato, who said that knowledge is virtue and must be available to all. The whole philosophy of Quran is based upon educating mankind, so that one can differentiate between good and evil.

Once the Prophet was asked what was the last thing he would do if he had only the last moment to live. He replied, *"Educate."*

5.3 The Scourge of Drug Trials

Research is inevitable for the progress of medicine. Drug trials are a part and parcel of medical research. New cures have to be found not only for new diseases but also for the old but previously incurable diseases. Without drug trials, such progress is simply not possible. The pharmaceutical industry, in collaboration with physicians and scientists, is obliged to carry out such drug trials, and they deserve to be complimented for their efforts. Safety and efficacy of the drug is the objective in such human trials, which are often conducted under scrupulous and controlled conditions.

It is indeed a global practice that scientists carry out lots of research putting several years of toil, sweat and labour into developing a molecule, then testing and trying the substance in many different laboratory and controlled environments before developing a medical formula to be tried on animals. A Phase IV trial is finally carried out under strictly monitored conditions on human volunteers. This is the job of the clinical scientists who have to seek approval of ethics committees and go through several protocols all designed to ensure the safety of human subjects before actually launching an experiment on healthy volunteers. It is usually a long and tedious exercise and may take several years, even decades, from the development of initial chemical compound to actually launching the finished product for a physician's prescription.

It is minutely monitored by many parties involved in such an evolutionary process, including quality control scientists of the parent manufacturer. Financial and moral responsibilities could be boundless.

On 13 March 2006, such an exercise was carried out on eight young and healthy volunteers at a renowned Hospital in London.

The first clinical trials on healthy individuals was about to commence at Parxel's chemical pharmacology research unit. The drug under trial was labelled as TeGeneros TGN 1412, a pseudonym, until a formal trade name could be given. It was a T-cell antagonist which was supposed to open new vistas in the field of immunology.

Each volunteer was fully informed of the trial and its possible hazards before the introduction of the drug. Each person was also paid a sum of £2,000 in lieu of the services offered, a practice not uncommon in the field of pharmaceutical research. Whether it therefore became a paid job or a volunteering service remains debateable. Nevertheless, the human subjects were healthy and joined the trial of their free choice.

What followed next could not be called anything but a total disaster. The clinical trials went horribly wrong, and the healthy subjects became critically ill patients. Some even developed multiple organ failure, caused by unknown and totally inexplicable immunological response to TGN1412.

The news of the tragedy became the hottest news of the days and weeks to come. Hourly reports of their progress was displayed on nearly all channels on British TV, and regularly broadcasted on hourly news bulletins on the radio. The media frenzy was typical of news-hungry media, who are always eager to jump and grab anything exciting. The paparazzi do not have any colour code or moral boundaries.

For nearly 2 weeks, the subjects under trial remained in a critical condition, obviously receiving first-grade clinical care. No one knew how to undo the effects that they had developed, and the diverse reaction to the drug under trial so far unknown and unseen created unprecedented dilemmas for the clinicians and the scientists. Top clinical immunologists and physicians and a team of pharmacologists continued to pour over the evidence and vast numbers of investigative tests that these subjects were obviously exposed to. No stone was left unturned in saving their lives and enabling them to regain their former health.

Eventually the tide turned in favour of the patients, and they gradually began to recover, finally returning to normal after a long and hard battle against the most horrendous immunological adverse reaction that anyone had ever seen. Some have not quite fully regained their former status of immunity and resistance. The media informed us one of them had shown early signs of developing a malignancy, possibly a lymphoma.

All of them are now leading a normal though somewhat traumatised life, contemplating litigation and compensation from the parent pharmaceutical company, which has now ceased to exist.

Following this near fatality, new guidelines for research and clinical trials have been published by the competent authority. Obviously, more vigilance and better scrutiny to ensure the safety of human volunteers would be the objective of these reforms.

Investigations and inquiries are a part of the British system of probing into such mishaps. The causes and effects are always thoroughly searched and published in due time. Incident recording and monitoring is a way of life, which in turn prevents such incidents from happening, which may otherwise turn into accidents and tragedies.

The Medicines and Health care Products Regulation Agency conducted an investigation on the TGN1412 mishap, which has been questioned by the parties involved as well as by the *BMJ*.

An independent Expert Scientific group was hence appointed by the government 'to learn from the Parexel clinical trials incident' as reported in the *BMJ* of 5 August 2006.

The usual saga of denials and disapprovals continued in the press. The *BMJ* reported that the recommendations fall into several broad categories, 'Preclinical development that is both directive and consultative.' Evidence-based transition to testing in humans, more open regulatory and ethical review, including independent scientific expertise and most importantly the need for more transparency were required.

It is indeed heartening to note that the recommendations will ensure safety and transparency of any future clinical trials.

The clinical trials will undoubtedly continue to be carried out and it is also beyond a shadow of a doubt that once in a while adverse reactions and incidents will occur. But it is expected that an important lesson will be learnt from this tragedy. Safety of human volunteers is of paramount importance in all clinical trials. What is equally important is that constant and acute monitoring of any trials by regulatory bodies, who obviously would observe the code of ethics, based upon universal principles of autonomy and fair play, shall be genuinely carried out.

The Thalidomide tragedy is not yet forgotten and shall never be in the future either. One had to pay a visit to the Trafalgar square in London, where just in front of the National Art gallery and in the shadows of the Nelson column there was controversial new monument of a limbless mother erected as a constant reminder of the follies of mankind and its dire consequences. The Tuskegee trial of Alabama and many such tragic episodes in human history are for some to remember and reflect so as to avoid untoward happenings, some unforeseen, others deliberate and totally evil.

While the question of ethical values, monitoring and transparency of trials on human beings is quite obviously of paramount importance in the developed countries, I am not entirely sure if they have the same significance in practice in developing countries. There is a lot of lip service in many such nations, but little to show in terms of actual practicality.

I recall participating in an international debate on the ethical issues related to human clinical trials, held in a hotel at Karachi. A scientist from the UK, a renowned ethicist, mentioned several illustrations of abuse of the powers of the Western scientists in employing colleagues for developing countries. The agenda was mutually beneficial to the promoter and the collaborator, but not nearly enough to the human volunteers, who often enough were grateful to be given some form of treatment, at all!

One illustration of the abuse and misquoting of the data was mentioned by this visiting speaker, which was an eye-opener for the whole audience, some very senior professors of medicine, who simply couldn't believe what they were hearing. In a drug trial conducted in a western country on a geriatric population, the data was gathered and employed in favour of the drug in question to be prescribed to young, healthy, pregnant mothers in many Asian countries. Some of the subjects on whom the study was carried out were in their 80s and mighty sick; some even might have had near death experiences, only Lord knows. To use such data is not only unethical but quite contrary to the basic principles of medical practice.

The contrast between the health services in the Western nations and those in the developing countries is enormous. There is no shortage of talent, the knowledge or the skills in our nations, but the lack of resources make the services totally incomparable.

Malaysia is a developing nation which has excellent health services which can be taken as a model for many other developing nations. In Malaysia, medical ethics is not only highly developed but pioneering in many ways. They practice what they preach. Some of their teaching modules are simply far superior to many Western countries and some of their ablest physicians and surgeons can simply outclass any university hospital staff in UK, and yet they are extremely humble and modest in their approach.

In the Indo Pak subcontinent we have some fine examples of institutions where excellent care is provided and can be the role model for other institutions. I can only quote an example of the Sindh Institute of Urology and Transplantation in Karachi and the Shaukat Kahanum Hospital in Lahore as two charitable organizations that are beacons of light for many corrupt government institutions. All India Institute Delhi and the Postgraduate centre at Chandigarh are two examples amongst many in India where students from the US and UK are routinely dispatched for hands-on training. These institutions follow the strict guidelines of the final moral values that are universal and have no limits or bounds of faith or religion.

The greener pastures of England and Wales and the majestic lochs and hills of Scotland offer too much to resist in terms of natural beauty and charm. The lure of the West is far too much to discard or ignore. Therefore, if a Western company or an institution involves a group of scientists or a person with authority to carry out a collaborative research, it obviously becomes quite difficult if not impossible to refuse such an offer.

The majority of such programmes are genuine and highly productive, but some evil scientists may barge in, and exploit the situation to their personal advantage and for their personal glory, even at the cost of human lives. Such evil people have no religion or faith or values. They are far too greedy and far too self-centred and ego-istic to care about a thing in the world. To them, self-interest is supreme and a good enough cause to embark upon any exercise good or evil.

A fine illustration of a collaborative programme was the Basic Institute of Medical Sciences, established in the 1960s at the Jinnah Hospital in collaboration with an American university. Honest and ethical people controlled it and the result of it is visible anywhere in Pakistan. Despite the closure of the collaboration, the BMSI remains an excellent institution producing the finest scientists to the country. An institution whose foundations were laid upon honesty has continued to serve its people with integrity.

There are many similar illustrations of such collaborative programmes elsewhere in the developing nations. A British charity by the name of BRINOS has provided excellent medical and surgical care to the ENT patients of Nepal over many years. As a fallout of their effort, the young British trainees learn the art of Mastoid sur-gery under the watchful eyes of the masters. It is a fine illustration of symbiosis, though some die-hard moralists have shown concerns regarding the matters of con-sent. Such a mutual contract is quite commendable.

Since many developing nations are often enough hard pressed for drugs and equipment, collaborative programmes are to be recommended for the mutual benefit of both parties. Keeping the principles of distributive justice in view, it is mandatory that the benefits of such trials and research reach both the researcher and the population under study equitably and without any restraints. It is incumbent upon the rich and the famous to involve the developing nations in research activities so as to benefit mankind, but also to strictly observe the ethical code of clinical and research trials on human beings so clearly mandated by the WHO and other agencies.

Regrettably, the Tuskegee and similar experimental works continue to remind us that the mighty and the powerful do what they want and would not care to pay even lip service to what the critics say. Haven't we seen it too clearly in the recent debacles of Iraq and elsewhere? Socrates believed in the might of the right, but he would be sad to witness for himself that in today's world the might is right.

5.4   Human Embryonic Stem Cell Research

The latest expose' of advancement in stem cell research has opened yet another Pandora's Box in England.

Media frenzy is a customary phenomenon here. So much time is allocated to the current affairs and news channels that the media are always looking for excitement. Be it a war or famine, an act of God or an evil action of mankind, the media are ever ready to highlight the matter. Often enough the voracious appetite of the media for exciting news leads to disproportionate explosion of the news.

Stem cell research is one such subject. It is exciting, innovative and offers great hopes in the future. It is in its early days, deserving more favourable attention of both the media and the public.

BBC is considered to be the mother of the news and current affairs. Fondly, it is called the 'auntie' by the most popular and globally renowned journalists like John Simpson.

Amongst many channels of the BBC radio, Radio 4 is considered to be the most cerebral and influential channel, with commentators like John Humphreys at the helm of affairs. Last week, Radio-4 broadcasted a debate on the moral issues arising out of stem cell research in today's medicine. A number of renowned and respected ethicists and moralists, teachers and philosophers argued the issues for an hour. It revolved around the latest development in the field of stem cell research, which is mind-boggling, fascinating and highly intriguing.

The issue under discussion was the latest breakthrough, in which a stem cell from a female mouse has been harvested, and cultivated in vitro to transform into male spermatozoa.

The obvious fear of the experiment, indeed the fearful conclusion being that soon the male species will become useless and obviously redundant. The male sperm will henceforth be obtained through male or female stem cell and not the male gonads and the relevant reproductive system.

One elegant female speaker , who obviously was single and a great supporter, indeed, promoter of women's emancipation went on to claim that soon the world will make a man obsolete and totally redundant and the age of the solitary parenthood will be the new world order. How frightful indeed!

No doubt the dilemma the world shall then face is colossal; however, like any other new development, the present breakthrough in technology shall also go through the test of time and be either accepted or discarded.

From the viewpoint of an ethicist, it is interesting to note that the scientific developments in the field of reproductive medicine employing stem cell is opening new grounds, which have been hitherto unexplored.

Let us examine the problem from a religious as well as scientific point of view. God has created mankind in His image, so says the Bible. And the Quran gives clear mandates on the creation of Adam, describing him as the vice reagent on earth, knowledgeable and worthy of respect by the other creations of God such as *jinni*. Man is also granted superior status through the title of *Ashraf* ul *makhlooqat,* i.e., the most respected or revered amongst creations. He is granted by Allah the most important talent or an attribute in the form of Rationality and reasoning, i.e., *aql -o- idrak.*

Like all other living creatures, mankind is also granted by Allah instincts like curiosity, security, hunger, greed, emotions and fears. But one thing that clearly isolates him and makes him the Supreme Being is the attribute of *Aql –o Idrak.* More often than not these attributes control the action of mankind in day to day life; however, sometimes emotion overtakes reasoning and the result is usually unworthy. How true is the statement of Ali AS, that anger begins with stupidity and culminates in remorse.

Therefore, rationality and reasoning are the hallmarks of mankind and should dictate major as well as minor decisions in life.

In the current debate, also, the scientists are employing the highest of human attributes, i.e., intellect, to do something that may carry inherent benefits for mankind. If there is even an iota of goodness involved in an action, then it must be called good and not evil. So why the debate?

There are many grey zones that need further exploration before reaching a final decision in the matter at hand.

The research activities in the medical science are being constantly monitored and audited by competent authorities, particularly in the developed countries. However, the situation is far from satisfactory in the developing countries. In fact, one major concern of the ethicists is exactly this, that if a technology such as the one under discussion falls into the hands of rogue scientist, we may end up with disaster, the remedy for which may still be unavailable.

The first successful isolation and culture of the Human Embryonic stem cells was carried out in 1998. Since that fateful day scientific research into the possible application of stem cells in the cultivation or replenishment of diseased organs is rapidly progressing forward in leaps and bounds. It is expected that in a few years time, this innovative technology may help eradicate such nasty and crippling conditions as spinal degenerative diseases, Alzhiemer's, Parkinson's and even common ailments like diabetes and heart ailments.

Stem cell research into the development of sperm does not necessarily mean bad news. It certainly offers new hope to many such parents, who for one reason or the other may not be able to have a biological child. One such example is a father who may have oligospermia or azospermia for idiopathic reasons, and would want to have child from his own genetic material. Another illustration may be that of a male who may become sterile due to irradiator for testicular carcinoma, or following surgical or chemotherapeutic treatment for prostate, testicular or penile malignancy. Therefore, to discard a technology on the superficial basis of an argument that it is unethical because it fails to meet the strict criteria of the ethicist cannot be justified. In all matters of biological sciences, some leverage has to be given to the scientists and the technologists, or else the sciences will stagnate, suffocate and succumb to lack of progress.

I feel the proponents of the debate favouring the stem cell research must be complimented for their fine efforts, and opponents should be given the due credit for keeping a close watch on the if's and but's. Checks and balances are always good for a healthy progress.

The germ line cells are obtained from the embryos and a specific number of germ cell lines are now available. Their development in the lab may result in huge impact on human race as it may change the genetic code, gender, even the fundamental personality of an individual. No more are allowed to be developed by the regulating authority at present. However, somatic cell lines can be developed and are being developed each day.

The launching of Human Genome Project was indeed a major step forward in the field of modern medicine. Stem cells are the back bones of the Eugenics and currently evolving genetic engineering. These stem cells are the Pluripotent cells that have a short life in the earliest stage of embryonic development. They have an inherent capacity to develop into any form of cell, tissues or an organ derived from the three germinal layers, namely the Ectoderm, Mesoderm and the Endoderm. In Surat Al *Zumar*, (39:6) Allah ST addresses these layers as the layers of darkness which could be the abdomen, uterus and the amniotic sac or the Ectoderm, Mesoderm or an Endoderm.

The embryology is categorically described in the Noble Quran in Surat al -*Hajj* (22:5) and Al -*Mominoon* (23:14) describing the formation of a zygote, a morula, a blastula and finally the embryo. It has been discussed elsewhere in this book at some length. But suffice it to say that it is the Balsocytes that actually have the inherent capability of differentiating into three germinal layers mentioned above. The problem is how to harvest the stem cells. Until now, the main source was the embryo, or the umbilical cord, though newer technologies are under development to investigate other sources like the adult stem cells from skin, etc., or even preserve the integrity of the Trphobalsts and the Balsocytes while obtaining the stem cells.

Coming back to the question of employment of embryos, the ethical question of the rights of an embryo have been raised by some diehard moralists. Others have objected to the very basic use of embryos anyway. Some others label it as murder of the embryos.

The question that arises out of all these and other discussions is that of the concept of life and its origin. Some believe that even a sperm is alive. A cosmologist believes that an atom is alive; that is why it has its offspring like electron, proton, photon, neutron, Quartz and Bosons, etc. The believers of Entanglement theory proclaim that for each atom on this planet, there is its identical pair elsewhere in the universe doing exactly the same as the atom at this earthly abode. Arguments may go on unabated. Suffice it to say that the life indeed begins at the time of conception. But emolument is different. The life before ensoulment is a vegetative state like the life in trees, plants herbs and shrubs. It is devoid of feeling, pain or suffering. It has no brain. So, really, what it means is that the life is nothing but an activity of brain. Is the brain the same as the mind or are they two different entities? The brain is in fact an anatomical organ just like an eye or lips or tongue, but mind is akin to the soul in the human body. Therefore, one may ask , when we call someone brain-dead, does it mean an organic death of the brain, the anatomical organ or the mind which carried all the feelings, emotions and memories? Philosophers have discussed and debated the difference between the mind and brain for centuries and left us no wiser, I dare say!

The other monotheistic religions have their own views. For instance, Christianity believes that life begins at the time of conception, but Judaism believes that life begins at the 40th day after conception. Islamic belief based upon Prophet Mohammed SA's tradition is that ensoulment occurs between the 40th and 120th day after conception. So the embryo is not really alive before these time frames. Or is it?

The embryo may be alive, but it has only a vegetative life. No mandate is available in Quran regarding the time of ensoulment of an embryo. When the prophet was asked about *Ruh*, he was advised by Allah ST to answer to the question that *Ruh* is an *AmreRabbi,* i.e., an act of God.

There are traditional *ahadith* that claim that the prophet said that the emolument occurs on the 120th day after conception. So can the embryos obtained from aborted foetuses be employed for the stem cells? Yes should be the answer, as such an embryo will be wasted away any way. If it can help regenerate a tissue or an organ, then its purpose will be well served.

One other way that the embryos are obtained is through the IVF technology. Extra zygotes are fertilised to avail of the embryos through cultivation and harvesting of the stem cells from them. The practice is vehemently condemned by moralists as it implies that useful technology like IVF is being abused for an ulterior motive. The Muslim moralists condemn it strongly and say that any extra embryo obtained in the process of IVF must be carefully and respectfully destroyed. But should it be? That is where *Ijtehad* needs to come in. Once again the fundamental principle of Islam comes into play. If a process can help alleviate the pain and suffering of a human being or indeed save a life by developing an organ that is so damaged as to cause death of the owner, than surely an extra embryo may be used for the harvesting of the stem cells. And that is a million dollar question that we must attempt to answer until a substitute for the employment of the embryonic stem cells is achieved.

Senaie has performed an in-depth analysis of the stem cell research in the Islamic Republic of Iran. The *Marjahe Taqleed Ayataullah ul-uzma* Ali *Khamanei* took a very bold and pragmatic approach and granted his approval to carry out stem cell research in Iran, which is a role model for other Muslim countries. As we know, these cells are Pluripotent cells and have an inherent capability of developing into any of the three germinal layers, i.e., ectoderm, mesoderm or the endoderm, which in turn are the precursor of the development of bodily tissues and organs. The ectoderm produces the skin and its appendages, as well as the nervous system. The mesoderm results in the development of the muscles and musculoskeletal system and the endoderm forms the internal organs and their lining membranes, etc. A student of anatomy can seek further information from the text books in this regard. For our purposes, suffice it to say that the Stem cell is the mother of all tissues, organs and systems of a human body.

These stem cells can be harvested in only a few specific ways. Their major source so far is the embryo, in its early stages of development. The trophocytes are responsible for the formation of stem cells and they have only a limited time period, during which they can germinate into any form of tissues, predetermined by the DNA code.

Stem cells appear to be the future of science. Millions of dollars are currently being spent by the NIH in USA as in many other countries to develop the art of harvesting these cells in easier ways and to cultivate the bodily organs in vitro or in vivo. Much success has already been experienced in this respect. The famous picture of an ear grown on the back of a mouse in a lab is a reminder that uncontrolled experimentation can be a joke for some and agony for others. One has to remember that scientists must not be allowed to play God or their haughtiness may lead them to do serious damage to its victims. Allah has time and again reminded us in the noble Quran to refrain from being haughty and arrogant as mentioned in the *Surae Luqman* when the wise man teaches his son to walk softly on earth and to speak softly to the people as neither a hard walk can break the earth open, nor can the haughtiness make you touch the stars in the sky above. *Luqman* goes on to remind his son that the worst of the sounds is that of braying of an ass.

Stem cells can be obtained from the embryos as well as from the blood in the umbilical cord. The ethical debate surrounding the stem cells is not about the employment of stem cells for research purposes, but the way these cells are obtained. Obviously, obtaining them from an umbilical cord with mother's permission may not be an issue, but when the embryos are created in a Petri dish for this very purpose, then the ethical boundaries are crossed. IVF is a common practice nowadays. The embryos may be deliberately cultivated for providing the stem cells would be unethical, but to employ those spare embryos that have not been used for the IVF or cryopreserved for future cycles may be employed for stem cell harvesting. But even this practice is objectionable in the eyes of a few die-hard pro-life activists, who claim that the embryos in a Petri dish have a right to live and killing them by availing the stem cells is akin to a murder!

The question, once again, crops up as to the origin of life. Is an embryo alive? There are scientists who believe that since every atom of life is alive, even a cell is alive, let alone an embryo which is a living example of life itself.

Therefore, if the embryo is alive, one cannot destroy it deliberately or intentionally; however, the fatwa in this regard says that extra embryos obtained through IVF may 'be left to perish' on their own. The argument here is that if these embryos are going to perish any way, why not put them to good use, obviously with the informed consent of the parents. After all, Islam is a religion of *Falah*, and what better service to mankind than to save a life or indeed many lives through the stem cell technology.

Stem cell technology is rapidly advancing. Already scientists have found alternate means of obtaining stem cells, from skin, etc. Soon the debate on the role of embryonic stem cell will be over and done with, but until that time, the moral dilemma continues to baffle scientists.

In 1989, the Islamic Organisation of Medical Sciences (IOMS) investigated the role of spare embryos obtained in the process of IVF. The special committee set up for this task later on issued a statement authorising the use of frozen embryos for scientific research within the boundaries of Islamic law. This decision was later on investigated by the Islamic *Fiqh* association, also in 1990. They remained undecided.

In 2003, the World Islamic Jurisprudence Council held a meeting in Mecca and allowed the use of stem cells as long as the source of these cells was permissible under the Islamic law. That included the use of adult consented umbilical cord blood, placenta, or extra fertilised eggs following an IVF cycle. Any form of intentional abortion to obtain stem cells was not allowed at all [14].

Very recently, experts have advised us that life actually begins at 3 weeks of gestation. The scientific view of the beginning of actual life and not simply the vegetative life is defined as the development of the primitive neurogenic streak at the end of the third embryonic week, which is the precursor of the future nervous system. After this stage only one single individual can evolve into mature foetus (*Larejani*, *Zahedi* quoting *Blaint* 2001 in www.intechopan.com).

In view of what has become a reality concerning the possibility to store non-fertilized oocytes for later use, it is necessary to restrict the number of fertilised eggs to the necessary number required for single treatment so as to avoid a surplus of fertilised eggs. If these surplus fertilised eggs are found to be present, they should be left without a medical help so that the surplus eggs may die naturally [15].

The fear of abuse of these frozen embryos may be the underlying cause of this mandate. But the argument may be made in favour of saving these embryos and employing them for the welfare of mankind, obviously under strict ethical framework. The argument that leaving these embryos unaided, i.e., left to perish, is akin to killing the embryos is stated in metaphorical rather than practical terms. Besides, if that is the case, then one is agreeing with the concept that IVF-produced embryos, as indeed all other embryos, are alive.

Senaie has very capably summed up the question of sanctity of embryonic life: 'the moral significance of the early embryos therefore remains at the centre of the controversy associated with permission to use it, while its destruction for the purpose of the harvesting stem cells is incompatible with the notion of embryonic sanctity and respect for the pre-implantation embryo.'

It is not difficult to justify the use of spare embryos based upon the principle of '*maslahah*' which basically means 'for public good,' i.e., a matter or an act that is useful to mankind and is in the best public interest (*darura*). It should overrule the prohibition. Furthermore, the Muslim scientist is reminded of another important moral principle called '*fard kifayah*', which allows conducting scientific research with the clear intentions and goals to benefit mankind. Therefore, based upon these arguments embryonic stem cell research may be morally correct.

The reader is referred to the guiding principles on research on Gametes and embryos in Iran dated 2005, available on the internet. It allows the use of surplus embryos below 14 days for research, including destruction of the embryos. It also prohibits production of embryos for research purposes to do any form of eugenics hybridisation, etc.

In the United States, the National Bioethics Advisory Committee (NBAC) disallows the use of extra embryos for research purposes. In the UK, the Fertilisation and Embryology Authority allows it for research but disallows its implantation.

In March 2013 a magazine called the 'Heart Matters' , published by the British Heart Foundation, published that Cambridge University has successfully carried out a scientific discovery through which they have developed stem cells called Induced Pluripotent Cells (IPC) from a blood sample. These cells can then be induced to grow any tissue or an organ. It is therefore expected that the invasive procedure as well as harvesting stem cells from an embryo or the umbilical cord may soon be an outdated debate, only to be found in the archives.

The Islamic world stretches from the farthest corners of China across the ancient lands of the central Asia through Eastern Europe and formerly domains of the Ottoman Empire, the Middle East. It stretches into the Western Europe across the Atlantic to Northern America and then via California back to the East, the land of the rising sun, Indonesia and neighbouring islands. It has a huge following of more than one and a quarter billion human beings. They all have one faith but many ways of observing the code of ethics as dictated by their cultures, norms and habits. The important factor is that despite the diversity, every one follows the dictates of the final testament and the last prophet. And that is such a unifying point for the Muslim Ummah. The world is perplexed to witness the harmony and unity of the global gathering at Arafat; one wonders why this can't continue to happen the rest of the days in the year!

5.5   Can Stem Cell Research Offer New Hope to the Deaf?

The Burden of Disability (BOD) caused by deafness is quite aptly acknowledged by world authorities like the WHO. Its impact is even worse due to the deficiencies of social and rehabilitative services in countries like Pakistan. Most deaf people live a life of silent solitude.

The data on the incidence and prevalence of deafness in Pakistan is grossly deficient. Suffice it to say the epidemiological studies are the least of our priorities in health care. However, sooner rather than later we have to carry out scientific population-based studies to establish our values of deafness and hearing impairment. Until then, we must rely on the published local and regional data.

Genetic deafness is a global problem. Many of its causes are syndromal or asyndromal. We can hardly cure or eradicate these conditions. A ray of hope, however, now seems to shine through for many such conditions.

Genetic engineering is one of the marvels of this century. But more significantly it is stem cell research which is making the real breakthrough. It is opening new vistas not only for replacement or regeneration of tissues and organs but also for replacing defective genes. It offers new hope to scientists to cure and eradicate many genetically determined conditions.

So what is a stem cell? And how relevant it is to us, in eradicating deafness?

Earlier embryologists employed the terminology of 'Totipotent cells' for the mother cells capable of differentiating into three germinal layers, eventually developing into various tissues and organs. The terminology of Stem cell is a more recent nomenclature.

Modern embryologists inform us that there are different types of stem cells. For instance, the multipotent cells are capable of developing into specialised tissues, but are devoid of an unlimited capacity to renew their growth. The Pluripotent cells, on the other hand, have the capacity to develop into any type of cell. If derived from an embryo these cells are aptly called the embryonic stem cells (ES) However, if they are harvested from the primordial germ cells in foetus, they are labelled as the 'Embryonic Germ Cells.'

Recent research suggests that the stem cells can be engineered to grow the muscle cells, nerve cells, cardiac cells, haematological cells, etc. Recently, they have been stimulated to develop a complete liver cell.

British Scientists have lately made an historical breakthrough. They have employed adult skin cells for developing stem cells for research purposes. The initial experiments are being carried out in monkeys, and if successful, one may be quite close to cultivating the human organs without involving the stem cells from an embryo or the placenta. And that would indeed be major achievement in the field of medicine.

So how can this knowledge help medicine? Geneticists are over the moon these days, as each day brings in new hope for many hitherto unsolved puzzles. They are already claiming the benefits in Alzheimer's disease and in diabetes. It is envisaged that soon each organ may be grown in vitro to replace defective organ, thus overcoming the shortage of organs required for replacements.

Prenatal selection through genetic screening has now become a reality. Prenatal diagnosis (PND), which is currently widespreadin the UK, is designed to detect pregnancies affected by such conditions as Down's syndrome, a known cause of hearing impairment, and spina bifida. For the last couple of decades a technology called Pre-implantation genetic diagnosis (PGD) has been developed which allows the scientist to test the embryos produced in vitro through IVF, for screening out any genetic disorders before implantation. Parents can now choose the sex of the embryo, through a similar but intricate and expensive scheme.

Professor Robert Winston is a renowned scientist. In a recent series on British television, he demonstrated the various aspects of PND and PGD, with the stories of success as well failure. Ethical debate aside, undoubtedly these are exciting times for the research scientists.

Since stem cells are now being employed to substitute a defective gene or grow a fresh cell, tissue or an organ, we can hope that soon a Cochlea will also be grown in vitro and may be substituted to replace a defective one. That age may still be far way, but at least Pre Implantation Genetic Diagnosis offers a definite hope for such families who have many deaf children and are desperately hoping to have a child with normal hearing. Through gene therapy, surely many defective genes can now be replaced with healthy ones. There are many new openings in the field of genetic deafness that are still waiting to be explored, as the latest technology in the field of macular degeneration has seen the light of a new day through the growth of retinal cells in vivo.

Deafness has always remained at the bottom end of the scale of priorities, especially when compared with blindness and other handicaps; but we surely hope to benefit with the later day research, if we keep our ears and eyes open. Once they have eradicated the more salient causes of human misery, surely they would consider growing a crop of healthy Cochleae too.

Technology is perpetually in search of eradicating disease. The race is akin to the hare and hunter chase. Its application may be motivated by financial rewards but the fall outs benefit mankind in many ways. Cochlear implants are a good example to quote.

In the future, organ replacement or replenishment will indeed change medicine for many such conditions that have remained elusive for a very long time. Total deafness is only a minor example. Just imagine the improvement in the quality of life of patients with motor neurone disease or paraplegia. Exciting times lie ahead.

In the *Journal of Laryngology and Otolgy*, in June 2012 an article claims that the use of embryos as the source of stem cells may be already over. Fresh crops of stem cells are now being harvested by genetic reprogramming of human or mouse somatic cells to induce pluripotent stem cells production. These induced pluripotent stem cells have a developmental potency comparable to that of embryonic stem cells, with the ability to generate all functional embryonic cell types, including neutrons, cardiomyocytes and hematopoietic cells.

This latest technology of reprogramming of adult mouse or human tissues to generate laboratory-induced pluripotent stem cells or indeed the human tissues has certainly bypassed the ethical debate of employing the human embryos for obtaining stem cells, but also opened a huge number of possibilities for generating adequate numbers of patient-specific pluripotent stem cells to cultivate, replace, or replenish defective human organs.

Stem cells are now being used experimentally for regeneration of hair cells, auditory neutrons, neural progenitor cells, immortalised auditory neuroblast cells, etc. It is envisaged that sensorineural hearing loss caused by degeneration or total atrophy of these cells may not remain an incurable condition in a few years. InshaAllah, soon one may see the replacement of a dead Cochlea with a lab-cultivated Cochlea.

If they can harvest and grow a pinna on the back of a mouse, they can certainly cultivate one in the ear also, replenishing the dead Cochlea!

In fact, heart, liver, spinal cord, or any other organ may be developed by the genetic scientists for replenishment of a damaged organ, just as a legend called Ghalib predicted [16].

Mirza Ghalib was a true visionary. In the nineteenth century he wrote:

Ley aiangey bazzar sey ja kar dlio jan aur.

The time is not far when that just might happen.

Allah has commanded mankind to explore, search and find the answers. Man has been granted the status of the most supreme creature of Allah ST, on the basis of *Ilm* o *aql*. If he possesses neither or he does not endeavour to employ those attributes, does he deserve to remain so, I wonder?

5.6   Cybrid: A Cross Between Humans and Animals: Is Science Going Mad!

An unprecedented explosion of science and technology has resulted in daredevil acts currently being performed by scientists in this early part of the twenty-first century. Surely enough, the author of *Brave New World*, Audrey Hexly, could not have imagined that such fictional episodes could indeed come to see the light of day. He did write about creating babies in a test tube outside the mother's womb, but even he did not conceive an idea as repulsive as the one under discussion. He wrote about the indigenous creation of mankind through ingenious technology, but even he could not conceive a repulsive idea as this one where a Cybrid is to be created through a cross between mankind and an animal.

No doubt the stem cell research is a major breakthrough in the chain of events to come. It is here to stay, and given the control and monitoring, it should result in the achievement of miraculous cures for diseases like Alzheimer's and the motor neurone diseases, etc. But the idea of creating a half human-half animal embryo to harvest stem cells for research is not only revolting but satanic in nature.

Indeed it is a moral issue, and must thoroughly be evaluated in the light of the principles of medical ethic, lest it may be said by the cynic that science has always been ridiculed by the church and the society.

No doubt scientists are quite capable of making decisions based upon their knowledge of moral values and bioethics. But it is also true that often enough such principles have been circumvented, if not violated, by scientists in the past. Hence, moral issues should not be left entirely to the discretion of the scientists.

One reason for this suggestion is that scientists in general and doctors in particular are less than ideal in their level of the knowledge of philosophy of ethics and morality. In fact, until recently and perhaps even today, the concept of teaching medical ethics in medical schools and universities producing scientists is rather alien and not totally welcome.

Physicians in general, have limited knowledge of humanities. One wonders at the importance of teaching botany and zoology to physicians at higher secondary schools when they should be taught humanities, mathematics and at least the fundamentals of philosophy. After all, in today's world of holistic medicine, the knowledge of one's attitude, psychology and etiquette, economics, etc., would be more helpful than the knowledge of binary fission in a perennial herb or indeed the reproductive method of a frog.

The world has changed. The authoritative and paternalistic role of the physician has gone and so has the mutual faith and trust between the doctor and patient, all the more reason that curriculum be revised to match the needs of the time.

Recently, I read an old book on medical history. It highlights the importance of inculcating the principles of ethics in a pupil of science and medicine. A famous Arab physician of the bygone days is known to have said, "Medicine is not science but a fine art that involves in-depth knowledge of human body as well as the soul. Medicine is also not a science alone but a form of art as it deals with living human beings and not objects sitting on a cornice."

Many ethical dilemmas can be solved if only we would indulge into the moral aspects of an act and evaluate it critically. The philosophy of good and evil is as ancient as mankind itself. The issues which are good and those that are clearly evil do not pose the problem; it is the grey zones in between these two parameters that cause dilemmas and deserve investigation.

The second reason for not leaving the moral decisions entirely to scientists is that many of them have an ego problem. To take up an impossible task or accept a challenge under the oddest of circumstances is almost a second nature to most professionals. For instance, in the legal profession, the top lawyers are those who have earned the reputation of accepting the seemingly impossible cases in trials and won them. There is a kind of a macho culture seen in all professionals, a form of ego that seems to flourish across the ranks of most professionals.

I recall that a globally famous American head and neck surgeon dedicated his very renowned book to the "brave and the bold head and neck surgeon who ventures into the unknown territory, where no one has gone before." The statement is self-explanatory and invites young people to take up the challenge so as to beat the others. It is a commendable thing to be able to accept the challenge and meet it with open arms, albeit without crossing into the territories which may question the ethical value of such an act.

Experience tells us that those who have indulged in such heroic ventures do feel 'special' and 'privileged,' walking tall amongst peers and colleagues with a sense of pride and ego. *'To be the first amongst equals'* is a sense that was duly depicted by the famous writer and story teller Jeffrey Archer in his book by the same name. The sense of achievement, accomplishment and glory provide some people more thrill than a shot of adrenaline.

Therefore, if you have total authority over all matters of science and research, you may end up playing God, which is exactly the reason that checks and balances are absolutely vital in all matters of research and scientific advances.

The scientist without moral controls may venture into the heroic adventure, which may harm humanity more than it should benefit. And we must remember that one of the pillars of bioethics is 'Do no harm.' Wandering into the unknown territories carries the risk of falling into abysmal pits, though often enough, as the saying goes, the Fortune favours the brave and the bold; he may come out unscathed indeed as knight with the shining armour! Who knows!

Human ego is a strange emotion. It can be called an attribute, and perhaps not even a good entity; nevertheless, it exists to a lesser or higher degree in all human beings. Those who can curb it, control it, or get rid of it often are the heroes of the legendary tales and folklore.

All philosophies and definitely religions abhor and condemn ego and egoistic tendencies in their followers. Islam has clearly warned against it and the day-to-day example of curbing ego and inculcation humility is the daily salat. One bows and prostrates five times a day and more to the Almighty Allah, to whom alone belongs the glory and the majesty. Islam commends humility, total submission and observance of the rules laid down by its teachings. It clearly condemns haughtiness, ego, vanity and similar evil sentiments like anger, jealousy, prejudice and pride.

The present debate on the development of Cybrid is outrageous and evil. Whatever benefits it may claim to achieve through its development cannot outweigh the harm it shall bring to mankind.

Man is created by the almighty as the *Ashraful makhlooqat* and, according to the Bible, in the image of God. To degrade him to the beastly level would bring great harm to human dignity, self-respect and the fundamental human rights. Regrettably, though, there would always be some diehard supporters, who may commend the creation of a Cybrid on the basis of 'Consequentionalism'.

And that is the problem I have with the theories of Utilitarianisms. To say that the creation of Cybrid is a minor evil, which may result in bringing about happiness to a great mass of people, simply appears hollow and unworthy.

To me the simple fact is that the abuse of knowledge and technology is by itself an evil act. One simply cannot justify the evil means to achieve an end —however good it may be! And that too appears to be illusionary and not beyond the shadow of doubt on its final outcome.

On 5 September 2007, the British authorities, dealing with the highly controversial subject of developing a Cybrid in UK, has granted permission to the scientists to go ahead with the project. Of course the ethics experts as well as the religious authorities are deeply concerned and strongly object to such an act of mankind interfering with mankind, but in this day and age of total and often unquestionable autonomy, who cares?

In March 2008, scientists in the UK successfully developed the techniques for developing a Hybrid. It is a cross between a human's genetic material and an animal's DNA sample. The intentions are to develop an embryo in a lab, with a view to harvest its stem cells. The Hybrid embryo will be destroyed within 14 days of its creation. The scientists hope that the source of stem cells thus created shall be able to supply them with inexhaustible supplies for cultivation of various tissues. It is claimed that the Pluripotent cells will be the nascent source of growing such tissues as the nerves, blood vessels, heart muscle, kidneys, liver and so forth. Diseases like Alzheimer's, or motor neurone disease, which are so far incurable, will then be amenable to further research and eventually cure. The shortage of organ supply for transplantation will also be overcome through the harvesting of such organs in the lab.

The ethical debate against such an act is currently raging across the land. The Catholic cardinals have seriously condemned such an act and have launched a campaign against the government's proposal of approving the bill in Parliament. In fact, three Catholic cabinet ministers are under huge pressure to resign if the bill goes through. There is a growing demand on the part of the general public to allow a free vote in the House of Commons in favour or against the bill to legalise the hybridisation of animal/human cross.

Whether such a product would be helpful in procuring sufficient quantities of stem cells or not remains to be seen. The fundamental thought is good and may be supported on the basis of utilitarianism, but there is a lot to be said against the very nature of experimentation which raises many ethical questions. The principle of interfering with the process of the creation of life is certainly against the religious teachings, and strictly to be condemned by Islam. The bishops and cardinals in the UK are calling it an ungodly act of the proportions of the infamous Dr. Frankenstein.

Science is not against religion, nor indeed against the established norms. On the contrary, it proves the divine revelations through the evidence of scientific research. After all, mankind is commanded to look deep into the oceans and far beyond the horizons to search for truth. Allama Iqbal wrote:

Khol anakh falak dekh shafaq dekh saman dekh; mashriq say ubaharateyhuiay sooraj ko zara dekh.

Obviously his implications were not just confined to advising his readers to investigate the wonderful creations of Almighty, but also a warning to the West that the East may have yet to offer much to the world.

United Kingdom is now a country with many faiths and many religions, but the majority of Britons shun any form of religious faith, and do call themselves nonbelievers. One can see the evidence of their lack of faith in many dilapidated churches across the length and breadth of the country. However, in the matters of interference with the divine creations, most abhor the notion, and do look at the scientific developments like the hybrid and the cybrid with some disapproval.

A debate held by a national TV a few weeks ago duly highlighted the importance of such a research that would enable the scientists to untie the intricacies of nature and perhaps find a cure for many a ghastly disease like the motor neurone disease, etc. Most debaters argued in favour of such research to be carried out, but many objected to the development of a Hybrid. Many of such respondents were either members of clergy or supporters of human rights. By and large, most Britons keenly await a breakthrough in the cure of such disabling and miserable conditions. After all, in a society in which family support is limited and disabilities a liability, few would complain if an answer to the dilemma of life-time care of the disabled can be found.

To conclude, one might simply remind mankind that the ultimate power of the creation of life and control over death rests with Allah ST, who has granted Ilm o Idrak to mankind to experiment and explore. But the noble Quran reminds us time and again that mankind is produced out of the meanest bodily fluids, and the meanest of bodily organs, and should not forget his origins by attempting to act God-like. But as the Quran repeatedly reminds us, mankind is arrogant, selfish, self-centred, egoistic, hasty and full of blunders. So let no one attempt to fiddle with nature, lest a fictional Frankenstein may become a reality. And we all know the fate of that infamous Dr. Frankenstein. So be warned, o developer of the Cybrid.

Eugenics is one way of playing with the genes. There are two other modalities of gene manipulation. They are Germ line therapy and Somatic cell therapy; all these methods are now available and can be utilised for an ulterior motive.

One sincerely hopes that the conscience of a lonely scientist wanting to teach the world a lesson or two overpowers his evil desires and makes him do an act of goodness. After all, science is meant to bring joy, not grief, to mankind!

5.7   Medical Ethics: The Third Dimension of Medical Education

Is there a need for formal teaching of medical ethics? Well, until a couple of decades ago, perhaps the answer was no. But with the ever-emerging dilemmas that a physician has to face in the twenty-first century, educators have already commenced upon a long and arduous journey of incorporating medical ethics as a regular subject. The ancient Flexner curriculum is all but replaced by contemporary syllabi, and didactic lectures partly replaced with interactive Problem-Based Learning.

Medical ethics is like a maze. It is dynamic and ever-changing. It is also heavily influenced by culture, religion, faith, customs, values, norms and traditions. In developing countries as well as developed nations, pluralism and multiculturalism are posing major challenges to ethicists. Faith and culture-sensitive issues have to be handled rather carefully. In order to prepare the modern physician for challenging times ahead, it is urged that medical ethics should be formally taught.

So let us discuss it.

Not a day goes by that we don't hear the story of an ethical controversy, a dilemma or a debate highlighting the singular importance of medical ethics in the life of today's practicing physician. And yet not many physicians are quite as prepared to face the challenges as they should be.

Currently, formidable changes are taking place in medical education. A variety of curricula and syllabi are under trials in the universities in developed as well as developing countries. Obviously, the need for change has been felt by the pundits, due to shortcomings in the traditional curricula, designed by Flexner [1] in the early twentieth century.

Flexner was no ordinary person. He was a visionary and a great American educator. In 1910, after witnessing rapid advances in medical education in Europe, particularly in Germany, he wrote a critique of the falling standards of medical education in America and recommended a thorough revision of the curriculum. His report proposed that any medical education must begin with a sound foundation of basic sciences, which should be followed by the teaching and training in the clinical subjects. Along with William Osler, the Regious Professor of Medicine in Oxford, Flexner and others gave testimony on medical education before a Royal Commission in London, strongly recommending that medical education be restructured on a university-based pattern [1]. The Flexner report therefore became the foundation stone of medical education in the West.

The Indian subcontinent acquired its *modus operandi* of medical education from the British. Good, bad or indifferent, it has served the region well. However, since medical education is an active process, it is only fair to expect changes to meet current needs. For the past decade or so, educators have been engaged in similar exercises to meet the challenges of the twenty-first century. Traditional lectures and demonstrations have been partly substituted with small-group discussion and Problem-Based Learning in some institutions. Difference of opinion on the PBL versus traditional methods continues to remain in the forefront. The argument is not over yet.

One such debate in educational circles revolves around the teaching of medical ethics at the undergraduate level. Many feel that the curriculum is already overloaded and should be saved from adding on even a proverbial straw. But others believe that ethical dilemmas faced by a modern physician demands that medical ethics should be formally taught.

Three well known components of medical education are *Knowledge*, *Skills* and *Attitude*. The teachers focus a great deal on knowledge, much on skills, but little on the attribute of attitude. So far, medical education has been two- dimensional. But time is now ripe to add the third dimension, which should give it much-needed depth and magnitude. It is the missing link between the science and the art of healing.

It is worth a note that the premedical education in the Indian subcontinent differs significantly from the UK and the United States. At least 4 years must be spent at an undergraduate level in the United States studying liberal arts before entering a medical school. The subjects may vary, but many undergraduate programmes would include Humanities, Philosophy, History and Communication Skills, etc. In contrast, such subjects are usually replaced by a 2-year course in Zoology, Botany and allied sciences in the pre-med programmes in most places in the Indian subcontinent. The importance of humanities and philosophy, in particular, in the life of a doctor can hardly be overemphasised. After all, medicine is more an art than a science! That is why it has always been described as the 'art of healing.'

It may not be taken for granted that once a medical student has memorised his textbooks and acquired the professional training, he would become an ethical doctor. Yes, he may be knowledgeable and skillful, but not necessarily ethical.

Many international institutions, such as the American College of Physicians, have advocated the teaching of medical ethics at an undergraduate level [2–5]. The General Medical Council of United Kingdom has strongly recommended the teaching of medical ethics at the undergraduate level in Britain [6] and opined that the students will gradually increase their knowledge of ethical ideas, as well as improve their ability to understand and analyze ethical problems.

In a recent study published in a renowned journal on medical education, the authors claim, "Complex and ethically challenging situations occur commonly in medical education. Appropriate educational provisions therefore requires medical educators to be equipped with knowledge and skills to engage with students with ethical concerns"[7]. This study involved three British medical schools. All of them used PBL method for incorporating ethics.

Medical ethics is as ancient as the medical profession itself. It has however, come into focus in the recent decades, particularly after the exposure of the atrocities of Nazi human experimentation in the Second World War. But the exposure of the horrors of Tuskegee trials, the Baby Fae experiment of xenografting, organ thefts at Alder Hay Children's hospital in Liverpool, the Bristol tragedy, rogue drug trials in the developing countries, the AZT episode, trading of organs for transplantation, and many similar stories have continued to keep the world focused on ethical issues. The most recent favourable ruling given by the British law Lords in the matter of assisted dying is yet another major ethical issue causing great controversy in the medical fraternity.

Medical ethics practiced in the Western nations is based upon the universal principles, aptly described by Beauchamp and Childress [8] in their famous book. They are Autonomy, Beneficence, Non -malificence and Justice, and are based upon the rules of human dignity, rights of self determination and personal decision making, providing the best health care while avoiding harm and equitable distribution of resources and benefits.

In most secular countries these ethical principles are taught and practiced per se. However, in some Muslim countries, such as Saudi Arabia, Iran, Malaysia, Kuwait, UAE, Egypt and Pakistan [9], the rules of ethics are taught and practiced as determined by the noble Quran and the Prophet's traditions, i.e., *Sunnah*. The most basic rule in Islam is to '*Enjoin what is just, and forbid what is wrong*' 30:17 [9], the practical form of which is defined as, '*Takrim, Wal Adl Wal Ehsaan' meaning, dignity, justice and benevenve.*

The principle of human dignity and respect for an individual and mankind in different meanings to different cultures. The impact of culture on medical ethics cannot be underestimated. In most developing nations, the autonomy may be collective, even surrogate, in nature as compared to the West, where everyone is unto himself, often enough a cause of many a misery that the people, and particularly the youth, of Britain and elsewhere in the West face these days. Limitless autonomy leads to erosion of authority, ending up in many disasters. A balance must be maintained between rights and obligations, duties and rewards; or else anarchy may ensue.

In a study conducted by Ypianger and Margolis [10] in an Arabic-speaking population of medical students in the UAE, the authors were able to establish that the students could identify medical ethics based on Western thought, despite the linguistic and cultural barriers. The study highlighted the importance of imparting knowledge of medical ethics to the undergraduate at an early stage of their curriculum.

Medical ethics may be taught in many ways. Whether it is longitudinally integrated or horizontally depends upon the syllabus adopted by the medical school. Experts in medical education like Harden [11] and others [12] have often talked about integrated teaching, which seems to make a lot of sense to the modern teacher in medicine.

Holistic approach is obviously the choice in today's set-up, and medical ethics almost imperceptibly fits into the system. One cannot but include the teaching of human values in modern medical education. Perhaps the best way is through the PBL method of small-group discussions. After an introductory talk, even a lecture, the medical students may be exposed to live ethical questions in a hospital setting or through simulated patients, skills labs, workshops or hypothetical situations.

British Medical Journal published an extremely useful article on this subject [13]. It defines the core curriculum designed by experts in the fields of medical ethics and law. They recommended it to be included in the curriculum for undergraduates in the British medical schools.

It is felt that the medical students of the twenty-first century must be formally taught at least the normative principles of medical ethics. They should also be exposed to some of the common ethical questions. Lectures, discourses, small-group discussions, simulated scenarios and the use of skills labs could be employed for this purpose.

5.7.1   Teaching Medical Ethics

The four basic instincts seen in all biological creatures are hunger, security, curiosity and companionship (*Ehtiaj, Tahafuz, Tajassus, Rifaqat*). All animal kingdoms from the measly paramecium to the illustrious mankind have these four instincts in common. But only man has been granted the supreme status by the Creator on the basis of *ilm*. Ilm teaches one the art of reasoning. It is the faculty of reasoning that grants mankind the status of *Ashraful Makhlooqat*, as he has been bestowed with the power to think and act rationally.

Rationality is what should determine the fundamental issues of good versus evil. If what appeals to your reasoning and your intellect is a rational decision, it is usually a good as against a decision made on the basis of animal instincts such as emotions.

But how do you educate the masses if you, yourself, are not fully informed. And that is a sad story of the medical profession all over the world.

Rapid advances in the knowledge and technology have left the framework of medical education rather devoid of the capability to face the new challenges.

The economists often use a term which is absolutely the right one for what I wish to convey: capacity building. Yes, the medical profession is deficient in capacity building. The medical schools have to stand up to do the task of capacity building to cope with the changing needs of the profession. The medical curriculum must be updated or else it become fossilised.

So how do you impart education to the medical students on such issues as the Organ Donation? Well, here is a model for your appraisal.

Selected Study Modules or Special Study Modules (SSM) are now routinely employed in many universities for incorporating such subjects which do not form the core curriculum. An SSM can be easily employed in any semester in the 5-year programme. Ideally, it may be introduced right at the beginning of the medical education. It can then be modified each year in a spiral fashion to add more queries, more problems and obviously, more solutions to the problem.

Here is a sample:

Lecture:

What is an organ donation?

What is a transplant?

What types of organs can be donated?

Why would the transplant be required?

Statistical data for organ transplants in a given country. Universal and global data.

Small group discussion: PBL

Learning issues, e.g., Modus operandi of the transplant team

Importance of informed consent,

The next of kin, issues and solutions.

Communication with the donor family.

Consultations with the recipient.

Renal transplant.

The procedure: its pros and cons.

Technical details … a lecture by the kidney surgeon, immunological aspects of the transplantation, a lecture by the Immunologist.

Graft repetition and therapeutics.

Experiences of the recipient from the relative.

Experiences of the cadaver organ recipient.

Issues related with the termination of life, brain death, religious and moral implications of organ harvesting.

Dilemmas, e.g., Organs bought and sold, smuggled or harvested illegally.

Should commercialization be legalised? After all, blood is bought and sold every day in many developing countries, and Ova and Sperm banks are all commercially available in many European countries.

Outcome measures on organ donation awareness, organ theft, transplantation methods, rejection and its prevention, organs for sale, financial and fringe benefit issues, organ harvesting from condemned prisoners of conscience. Human rights issues.

Unless medical ethics is taught at the undergraduate level, the future physician is bound to find it hard to match the demands of the times. Many ethically challenging questions are bound to crop up in his life, to which he may have no answer. The basic job of the undergraduate teaching in the medical schools and colleges is that of laying a firm and solid foundation for the building to be built upon. A weak foundation may not enable the final structure to be strong and solid enough to withstand the onslaught of weather.

It is about time that the medical educationists would begin granting due recognition to formal teaching of medical ethics. Whether it is in the form of conventional lectures or small group discussions is a matter of choice for the revenant authorities. Medical ethics is a fascinating subject that can be taught through PBL as well as didactic lectures and discourses. Problem-based learning was the modus operandi of the Greek great Socrates, when he posed problems and never gave an answer. When asked by the cynic, he'd simply say that giving an answer will discourage you to find your own way through the maze. Learn to solve the problem through active engagement. Is that not the fundamental rule of the PBL? Be actively involved, learn for yourself and be a lifelong learner.

What better way to impart education on medical ethics than following in the footsteps of that great Greek philosopher?

Once armed with core knowledge of medical ethics, the physician must then continue to expand his knowledge through daily encounter. And that indeed is the current concept of medical education in brief, i.e., 'core knowledge at undergraduate level and more knowledge throughout life'.

5.8   Good Medical Practice, Nebuchadnezzar and the Hammurabi Code

General medical council has just dispatched its latest guidebook on Good Medical Practice to all its members and registered doctors. Like all other GMC documents, it is full of practical guidelines inundated with experience and wisdom. In my view, its opening paragraph is probably the theme of this booklet.

It reminds us that patients need good doctors, who make their care the primary concern, and serve them with competence, knowledge and skills, establishing good relationships not just with the patients but also with the colleagues. They are also honest and trustworthy, acting with integrity and within the law [17].

This document further elaborates the various components of GMP, such as the Domain 1 discussing knowledge, skills and performance, recoding one's work clearly, accurately and legibly, etc.

The Domain 2 discusses various aspects of safety and quality, basically directed to comply with systems developed and applied to protect patients, such as regular audits, feedback, confidentiality, noting down adverse reactions, etc.

Under Domain 3, huge importance is given to communication skills, developing partnerships and teamwork. It does not surprise one at all that communication skills are given due importance as they should indeed be. A doctor with effective communication skills is far superior in managing patients than the one who may be equally competent, technically even better, but a poor communicator. Patient satisfaction is an art that we must learn, hone and apply in all our daily activities.

Inter-professional relationships are another salient feature of a good doctor. Mutual understanding, respect, collaboration and teamwork certainly yield far better results than solo performance. A lonely scientist may indeed be a risk to the art of healing, as indeed to research!

It is also essential that we know our limitations and transfer the patient to a more qualified or competent physician at an appropriate time, and arrange for safe transfer of patients between health care or social services, as the case may be.

In these days of rapidly expanding knowledge and technology, no one can master all that is available. Super-specialisation has raised a few eyebrows, but perhaps it has now become inevitable. Therefore, one must treat the patient in accordance with his personal competence and knowing his limitations. It is far safer and better to admit your faults and deficiencies than treat the patient wrongly under the false impression of knowing it all. One must therefore take a decision as to the referral to a suitable speciality or a physician, if one finds it genuinely out of his comfort zone.

This document reminds us of the fundamental attributes that all physicians must possess such as politeness, treating patients with dignity and maintaining their privacy, respecting their life choices and their faiths and beliefs, explaining to them whatever is involved in their management, including risks and benefits, etc. The fundamental normative principles of medical ethics are reiterated here.

Despite the poor condition of the NHS, and daily drama that unfolds on the TV, one must admit that the GMC keeps us all under control, on the straight path, vigilant and regularly updated on the matters that concern us most.

This document is certainly worth reading for practical application.

Alas, the developing world is still living in the Dark Ages, as far as the principles of Good Medical Practice are concerned. I say that with the tongue in the cheek, but it is sadly true. The following story will underline the reason of my concern, indeed, despair.

Only last week (March 2013), I was in Iraq, giving talks on medical education and medical ethics to the faculties of the universities of Kufa and Babylon. Some amazing discussion surfaced during one of the sessions; hence, I thought it is worthwhile reminding my readers of not just the philosophy of medical ethics but also the principles of Good Medical Practice.

Visiting Najaf, Kufa, and Karbala is always the best experience one faithful can ever have. These holy places are second only to Mecca and Medina, but perhaps more visitors are seen here on certain days like the Arbaeen, than even at Hajj. Besides, one obvious thing is the total devotion, humility, dedication, sorrowful eyes and hands raised in supplication with more utmost sincerity here than in the Arabian cities. In fact, I was a victim of pickpocketing during the *tawafe* Kaaba during my first Hajj! No such thing ever happens in Najaf and Karbala. The whole experience is so very fulfilling, spiritually cleansing and morally invigorating that you return from these places as a newborn. Allah pays instant attention to the supplications of His creatures, when you pray in *Masjide* Kufa, or Haram e Ali o Hussain o Abbas AS. I have ample evidence of that in my own life and also in the lives of many others who share my experience.

The new *Zareeh* of Imam Hussain AS is a humble tribute of the followers of Ahal albeit but what a piece of masterly craftsmanship it is; only one who sees it can endorse this observation. The construction of all the Harams is giving them a new flavour, texture and colour. The Imam Ali mosque adjacent to the *Zareeh* is a glittering crystal house. And the front wall of Imam Ali's mausoleum called the Babe Muslim Ibne Aqeel is covered with gold. It is massive in size and only a simple reflection of how Ali is loved by the *momineen*.

Haram e Abbas As is known to all of us as the place where all our difficulties and trials and tribulations are instantly resolved. At the Babe- Hawaaij, you just have to pray to Allah ST in the name of Abbas, Babe- Hawaiaj and your problems are solved. I have written a full essay on Faith and would reiterate my earlier statement that Faith is the fundamental brick of Islam. If there is no faith, your som o salat are no more than a ritual. You beg Allah in the Haramaein of Ali, or Hussain or Abbas AS, with total submission to His authority, and you would not return unfulfilled. That is a statement that I wish to record here. Except for death, which is the ultimate reality, all your dilemmas are solved here.

The tyrant Saddam had destroyed all these holy places, so after the liberation of Iraq, one can see huge developments going on in all these places. Harame Ali AS is expanding to double its present size, as we saw the foundations and piling works going on, close to the hotel where we stayed.

Masjide Kufa is a very special place. This is where the greatest jurist of mankind, Imam Ali, prayed and distributed justice as the Caliph of Islam. It is also the place where he was murdered by an enemy of Islam, the *mahrab* in which he was leading the morning prayers when slain is a place of worship for us all. The mosque is inundated with places where the Prophets had prayed during their era. It also has the place where the floods of Noah emerged. The mosque also has the mausoleums of Muslim Ibne Aqeel, the ambassador of Hussain AS to Kufa and Hani bin Urwah, who was an ardent supporter of Muslim, and Mukhtar -e- Saqfi, the avenger of the massacre of Hussain in Karbala.

A few yards from Masjide Kufa are the humble dwellings of Imam Ali, which is a place where the Imam lived during his caliphate. All these buildings are renovated since the fall of Saddam and are rapidly expanding to accommodate millions of pilgrims each year. The same applies to Karbala, where new apartment buildings are being erected as the last Arabaen saw the influx of 10–12 million people, which is bound to rise.

A visit to the ancient city of Babylon was a lesson in history as we saw the 5,000-year-old Ishtar gate, and the castle of Babylon built by Nebuchadnezzar (Bakhtnsar 2) and the relics of the hanging gardens of Babylon.

Hammurabi lived and ruled here and gave his famous code of conduct to the world. I often quote it in my lectures as the first written document on moral code. But as my guide and learned physician, Dr. Hasan, informed me, since it was written on a clay tablet, it did not live for long, as against the papyrus used by the Egyptians in their documents. I beg to differ with Hasan, as what Hammurabi gave mankind lives on, but what the Pharoes gave the world is dead and gone, papyrus or clay. It perished with the fall of Tunekhamen and Ramses the second. '*Fa aattebaru yaaulialabsar*: (Quran; Al *Hashr*,' 59:2.)

On the way back to Najaf, I also paid tributes to my ancestor Imam *Zaid* bin *Ali* who was mercilessly killed by Hajaj about a thousand years ago. He is now buried in a place close to Hilla and Kifl. It was an awesome experience!

Now to the actual conference. One is amazed to note that in the regions where all the great prophets and apostles lived and died, so much immorality and tyranny has existed over the millennia and continues to flourish even now. It is mind-boggling, to say the least.

At the conference, Dr. Hasan, an eminent physician and epidemiologist, exposed a few sad happenings that the Iraqi medical profession is currently experiencing. We were shocked to hear that the patients not only sue a doctor if the results were not up to the family's expectations, but the whole tribe would rise against the doctor and either ask for preposterous sums of money as compensation, or threaten him with his life or have him jailed. Hasan said that in March 2013, at least 11 physicians including obstetricians were behind bars and there was no way they could get out. On raising a question of medical protection, it was revealed that no such thing existed in Iraq, so either you heal the patient or be ready for trials and tribulations.

There is no medical association to raise the concerns of the profession or to support them in the hour of need. What is worse is that the surgeons have become so scare, that even routine surgery such as a hernia repair is now either deferred or the patient is given a somewhat scary picture of possible fatality so that he decides against surgery.

When enquired about any research, my learned Iraqi colleagues looked at me in the eye and raised a counter query. Will you conduct any form of research when your own life is at stake?

Obviously not. Research requires peace of mind, tranquility and resources. If you are desperate or indeed a desperado in this case, you think of only one thing, and that is your safe exit from such perilous situations. And that is one of the major reasons, my dear readers, that research has been so deficient in the developing countries. Insecurity and living from day to day do not permit you to think creatively or to apply your gray cells in any other venture than daily living, earning your bread, keeping low and just surviving.

Recently, I was reminded in an article of the sad story of the only Nobel Laurette of Pakistan, Professor Abdul Salam. The gentleman wanted to live in and serve his homeland. He was disqualified to teach maths in a college, a subject in which he had already established his name in the world of science. Instead, he was given the job of a physical instructor, as he was physically well built, and in view of the selectors, perfect for the job! He took the job, only to live and serve his people, but was eventually sacked as he was to be always lost in his own world, in other words more absorbed in his cerebral and brainy activities than the brawny functions for which he was initially hired.

Eventually he had to migrate to Italy, where he established a world-class centre on theoretical physics in Trieste, which continues to benefit the world, particularly the developing countries, long after the Nobel Laureate left this world.

Similarly, awful conditions in Iraq have crated a panic amongst the profession and all and sundry wait for an opportunity to migrate elsewhere. It is sad, as thus the developing nations lose their talent and manpower to the West, who benefit with the experienced doctors and hire them at lower wages than their contemporary local doctors or nurses, often making them do such jobs that they would otherwise never take up in their own lands, if the conditions were right. So there is a moral dilemma for the society in those regions who treat their elite like criminals and those Western countries who reap the rich harvest of mature and ripened fruit, and abuse them at their own discretion!

Hammurabi of Babylon left us a code written in 1792 BC. It is oft quoted, but totally forgotten by the people of his land. How ironic is that? Surely these doctors do their best to serve those whose ancestors may well have been the subjects of Hammurabi and Nebuchadnezzar, as best they can; but no physician can guarantee a cure despite his intense, dedicated and determined effort. And these people of the ancient lands of the prophets and saints continue to behave like the cave man and take revenge through might and tribe. How pathetic is that?

It was so sad to talk to many doctors during the lunch break, who look helplessly towards anyone who can help overcome this unfortunate situation. Of course, the visiting international faculty suggested a few items, like the formation of a medical council and inviting international medical defence companies to insure against malpractice suits, etc., but who should bell the cat? And that is a million-dollar question in the land of the King of Babylon, who gave us the code of conduct several millennia ago.

The GMC is beginning its long-awaited revalidation process this summer. One question that came up in the discussion at the Iraqi seminar was exactly this. It was asked by a physician in the audience, whether there was any method of revalidation or CME process that doctors in the UK must pursue. Well, CME has always been the backbone of the medical practice in the UK. Besides, we all must complete a certain number of courses each year to keep ourselves worthy of employment in the NHS, such as the BLS, ALS, or ATLS, etc. Furthermore, we must also attend courses on anger management, conflict resolution, handling the babies, children, the elderly and the disabled, fire safety, handling and transferring patients, etc. Similar mandatory courses are conducted in the United States, as we were duly informed by the visiting American faculty.

The new item on the agenda is the revalidation of all working doctors by the GMC. Some may prefer to retire if faced with an awkward challenge in the future years. But it is all in the best interest of the patients, and we as physicians are obliged under oath to give priority to the care of the patient, over and above all other elements.

One of the messages given by the latest GMC documents is that a physician must be proficient and ethical. That, indeed, is the underlying principle of Good Medical Practice.

References

1. Pritchard L (2006) Bordahge: following in Flexner's footsteps. Med Educ 40:193–194
2. American Medical Association (2006–2007) Code of medical ethics. Current opinions with annotations. AMA, Chicago
3. Andre J, Brody H, Fleck L, Thomson CL, Tomlinson T (2003) Ethics, professionalism, and humanities at Michigan state University College of humane medicine. Acad Med 78:968–972
4. Murray J (2003) The development of a medical humanities programme at Dalhousie university faculty of medicine. Nova Scotia, Canada. 1992–2003. Acad Med 78:1020–1023
5. Montgomery K, Chambers T, Reiffler DR (2003) Humanities, education at North-western University, Feinberg School of Medicine. Acad Med 78:958–962
6. General Medical Council (2003) 'Tomorrows doctor'. Recommendations on undergraduate medical education. GMC, London
7. Cordingley L, Hyde C, Peters S, Vernon B, Bundy C (2007) Undergraduate medical students; exposure to clinical ethics: a challenge to the development of professional behaviour? Med Educ 41:1202–1209
8. Beauchamp TL, Childress JF (1994) Principles of biomedical ethics, 4th edn. Oxford University Press, Oxford
9. Al Quran, Luqman 31:17
10. Ypinnager VA, Margolis SA (2004) Western medical ethics taught to junior medical students can cross cultural and linguistic boundaries. BMC Med Ethics 5:4
11. Harden RM (1998) Integrated teaching – what do we mean? A proposed taxonomy. Med Educ 32:216–217
12. Dahle LO, Brynhildsen J, Berbohn Falsberg M, Rundquist I, Hammer M (2002) Pros and cons of vertical integration between clinical medicine and basic sciences within a PBL undergraduate curriculum. Examples from Linkpong, Sweden. Med Teach 24:280–285
13. Teaching medical ethics: a model for the UK core curriculum. www.bmj.com/content/vol.316/issue7145
14. Fadel HE (2010) Developments in Stem cell research and therapeutic cloning: Islamic ethical positions. A review. Bioethics 26(3):1–8
15. Senaei M (2008) Human embryonic stem cell research in Iran. Indian J Ethics 4:181–184
16. Iberkewe TS, Ramma I et al (2012) Stem cells in sensorineural hearing loss management JLO. pp 652–656
17. Good medical practice. GMC, An official document if General Medical Council of Britain, 25 Mar 2013

Further Reading

www.Medicalethics.imamaimedics.org
www.imana.org
www.islamicmedicine.org
www.people.virginiaedu/~aas/htm
www.islam.org.za
http://actgnetwork.org

Reproductive Health

6.1 Contraception, Assisted Reproduction Technology, In Vitro Fertilization, Abortion, and Pre-Genetic Diagnosis

Savants raise many ethical questions on the innovative methods of assisted reproduction that are currently in vogue and gaining popularity each passing day. No doubt some of them are simply mind boggling, to say the least, whereas others do raise many questions in a critical mind. For those who have not been blessed with a child, these innovative technologies offer unprecedented hope and therefore bring good deeds, albeit not without causing some concern on ethical issues as a matter of course. For many others, some techniques are a good source of revenue – and here I mean the experts who employ these techniques, that is, physicians and scientists. What we need to examine in this chapter is, "Are they all ethical in their approach?"

An important conference on population planning was held in Cairo in 1994, which brought many ethical dilemmas into the limelight. Those were early days and medical ethics was just beginning to get a foothold in the Muslim world. The media took a special interest in highlighting the proceedings of this conference. An article was published in 'Daily Dawn Karachi,' the most famous English daily of Pakistan, founded by no less a person than the founder of Pakistan Mohammad Ali Jinnah himself, on the controversy of abortion. Is abortion legal or otherwise in the light of Islamic teachings?

The writer, Ahmed Afzaal [1], discussed the agenda of the Cairo conference namely (a) population planning, (b) gender roles, (c) sex education, and (d) abortion. He went on to discuss contraception at length. He argued that a fetus does not have a name or an identity. In other words, it has no persona, and is before the ensoulment – which, according to tradition, happens after 120 days of conception – so early Muslim scholars had allowed the individual abortion before this cutoff point.

This may not be true; we know that ultrasonography can detect the heartbeat of a fetus as early as 6 weeks of pregnancy. Hence, if life is considered to be related to a heartbeat, 120 days may not be the best cutoff point. In fact, recent scientific studies have identified and determined that the development of the primitive neural crest, the precursor of the future central nervous system, is the cutoff when an individual entity can be definitely determined, and that happens after 3 weeks of conception.

Afzaal had argued that the relevant Islamic tradition actually dealt with infusion of the spiritual soul (*ruh*), which is not identical to life. An embryo is alive at every stage of its development. The learned author argued that even the zygote and its precursors, ovum and sperm, are living cells. In fact, some ethicists differentiate between a vegetative life and a true life, believing that a fetus before ensoulment may have a vegetative life and not a true life.

We know that Islam clearly mandated against all forms of abortion, except if the mother's life is in serious danger. A council of doctors and ulema, as members of the ethics committee, may then decide in favor of an abortion.

In the spring of 2012, in a collection of articles edited by a celebrity, I read an extremely interesting article on the life of carbon. The author believed that carbon is alive and the first iota of carbon that came into being millions of years ago is still in circulation. He describes the life cycle of the carbon molecule in an immaculate fashion. When we die, the carbon molecule that forms the link in our DNA chain is released in a life form that is consumed by insects or goes into plants, which then consume it in the manufacture of the chlorophyll. The animals eat that plant or its products, and humans eat the animals, and the carbon molecule continues to remain active, alive, and building blocks in one form or another. We then breath out carbon dioxide, which is happily absorbed by the plants and trees, who release oxygen in exchange, which we breath in . . . and life goes on. So the first molecule that came into existence at the beginning of life has not died; it continues to live, albeit in different forms. How intriguing indeed!

Rumi, the famous Persian philosopher and poet *par excellence*, wrote a couplet that is included in an essay in Sect. 2.2. It is amazing to note that hundreds of years ago Rumi described an almost similar metamorphosis of the human life cycle. So the moral of the story is that if a carbon atom is alive, surely sperm and indeed an ovum would also be alive.

The creation of man is described in the Holy book in the *Ayat-e-Mubaraka*; "We have created man out of clay; then into a clot, then into a lump. We then created bones which we clothe with flesh. Then we transmute it into a new mode: blessed be then Allah the best of creators" [2].

This is, of course, the fundamental principle of the creation of mankind. In practical terms, it implies that a baby is born out of an interaction between male and female hormones. The fact that a fetus owes its creation to its parents does not necessarily entitle them to destroy it without valid reasons. The argument in favor of or against contraception is a time-honored debate.

6.1.1 Contraception

Contraception was practiced in the Arab culture before the dawn of Islam. The method used then was called 'Azl,' which means coitus interruptus. As to whether

Islam allows it is open to debate. Dr. Fazlu-Rehman, an eminent and visionary Muslim scholar of the not so distant past, had progressive thoughts and a pragmatic approach. He wrote aviyt the subject of contraception at length in his book *Health & Medicine in the Islamic tradition*.

He said that according to a reliable hadith, contraception was not banned by the Prophet. The Maulana quotes that a companion of the Prophet (PBUH) reported, "we used to practice coitus interruptus during the Prophet's lifetime and he knew about it while the *Quran* was also being revealed, but the *Quran* did not prohibit it."

The learned scholar therefore based his argument on logic and said that it seems plausible to hold that the pre-Islamic practice of contraception was allowed to stand by the Prophet (PBUH),without saying anything about it, although he could have banned it the Maulana argues "if the Prophet (PBUH) wished to do so. Since the Prophet kept quiet about it, it can be assumed that he did not have any objection to that practice". We know that the *Quran* has given a clear adage that the Prophet only commands under a *vahi* ordained by Allah. So if the prophet had any commands against the practice, and Allah wished to ban it, the prophet would have spoken.

Fazlu Rehman also quotes another Hadith in his book, saying that the prophet (PBUH) said that on the day of the resurrection, he would be proud of the numbers of his communities and portrays him as admonishing his followers, "so reproduce and increase in number." Of course we all know that in the *Khutbae-Hajjat-ul-Wida* (the last Hajj) the only Prophet condemned genocide most strongly and prohibited any form of violation of human life. Perhaps one might argue that in the ancient days, when wars waged between the tribes, or hands were in much demand to toil in the soil, or labor in the fields, or the care of cattle in the barns, a large progeny was a welcome – indeed a valuable – asset. In those days, a lot of men died during the wars and many children died of disease, hunger, and famine – as one witness in today's Afghanistan and elsewhere – so the need for contraception did not arise.

Dr. Rehman argues that many theologians and lawyers permitted contraception and some even abortion within the first 4 months of pregnancy – before the fetus is "infused with life". According to him such family planning was never adopted as a national policy – and was left to the individuals.

Continuing his commentary the Maulana states that different religious sects in Islam have viewed contraception differently. For instance the Zahiri School of thought – quite prevalent in the Arab Middle East, absolutely forbids contraception as well as abortion.

On the other hand according to the Maulana the 'Shafli' school allows contraception unconditionally to the husband who need not do it with the consent of the wife; for they contend that a wife even if she be free and not a slave has no automatic right to children but only a pleasure [3].

Not all followers of 'Shafai' School of thought however, ban the contraception or family planning. A great theologian and Shafai philosopher, indeed a Sufi Al-Ghazali believed that Allah has promised to provide 'Rizq' – 'sustenance' to each soul He has created. Therefore it is unbecoming upon a slave of God to violate the law ordained by his Master. However, Al-Ghazali argues "that if one fears that his children may suffer economic constraints, or if his wife may lose her health, or her looks – so that he might come to dislike her – a man should refrain from having children" [3].

In the pagan Arab world, a male child was considered to be a reason of great joy to the family, indeed an honor for the father, even the tribe. A female child was thought to bring disgrace, hence not considered worth allowing to live. Newborn baby girls were buried alive, we know through history. Islam prohibited such practice with strict and firm command. Indeed the Holy Prophet (PBUH) himself had no surviving son, as Ibrahim and Qasim both of his sons died in his life time. He simply loved his daughter Fatimat-ul-Zahra (SA) so much that he called her "*qurat-al-ain*" meaning "an apple of my eye" and "a part and parcel for his very flesh and soul."

Pakistani culture is an amalgam of the Arabo-Iranian as well as Indo-Harrapan civilisation. We have imbibed the best of both civilisations, but we have also acquired some of the worst traits. Hence, just like in the Pagan Arab world, "while a male child is a bonanza, a female child is a curse."

Thus if a mother has three consecutive daughters, she must continue her ordeal of pregnancies until a son is born – or else she must suffer disgrace, and humiliation at the hands of the inlaws. Islam does not allow such a practice at all. All creatures are born out of the will of God. And every soul has to return to Him.

The impulse to procreate is especially powerful in our rural areas. Notwithstanding the ultimate fate of the children the procession of the babies continues to move on at the very basic level of sustenance.

As discussed before the time period at which a fetus who has a vegetative life, changes into a live baby has been detailed at 120 days of pregnancy. It is at this time that the fetus becomes ensouled. It becomes a 'person' in legal terms. It has now inherited all traits of its parents, and indeed all those rights which its parents possess.

Genetic engineering, test tube babies, I.V.F. (in vitro fertilization) and other similar acts have undergone evaluation in the light of Islamic teachings.

6.1.2   Test Tube Babies

The concept of test tube babies has been evaluated by the Muslim jurists. In its early days, the only theological scholar who had discussed it is Dr. Fazlu-Rehman, who saw no fault with it, indeed actually found it a "welcome development," provided such a union was strictly between the genes of husband and wife. He did not consider it to be any form of interference with the God's work. Indeed he considered it to be akin to a process of a sapling cultivated under controlled conditions, and transferred to its proper place at an appropriate time. One must admit that Dr. Rehman was far ahead of his times. He spent most of his time in the USA. He was too pragmatic and progressive in thought and teaching to have survived in the dogmatic East!

No doubt it is an appealing argument, since the plight of childless couples cannot be estimated by all. And, if only, they can have the joy of having their own biological child, it would indeed mean giving a new lease on life to them. Adoption is not a natural way, nor is it a genuine substitute for one's flesh and blood. An adopted child is not a biological child of the adoptive parents. Besides he can neither inherit the name nor title as well as legal inheritance from the adoptive parents.

Some jurists argue that 'adoption' is discouraged in Islam just like the pre-Islamic custom of 'Zihar' – as a way of creating an unnatural relationship. It is not true. The famous illustration of Zaid, the adopted son of prophet is known to us all. In fact his name is mentioned in Quran with reference to his wife, whom he divorced. Prophet SA later on married her. Then there is the example of Mohammad Ibne Abi Bakar who was the adopted son of Ali AS. His mother married Ali AS after Abu Bakar passed away. Mohammad was as close to Ali AS his own biological children, but was always called Mohmmad ibne Abi Bakar. The living proof of Ali's love for Mohammad Abi Bakar is evident from the facts that Imam gave him huge administrative responsibilities during his caliphate, and married the daughters of Iranian King Yazd Jard who arrived as captives, to Hussain AS and Mohammad Bin Abi Bakar.

Islam came to reform the society, to improve the fate of the lot – and not to promote poverty or misery. Hence. keeping the very basic doctrine of '*ijtehad*' in view, it goes without saying that population control is the cry of the time.

Therefore one would argue that while routine abortion should not be condoned, contraception should be allowed as an individual choice – if not as a national policy.

IVF and assisted reproductive technology is now allowed in Islam, but not the third party gamete transfer, which remains controversial.

Some ethicists believe that the dilemmas begin before baby is conceived. In fact they believe that the parents' thoughts, pattern of life, their own upbringing, etc. influence the process of copulation and conception. It is therefore said that a child is indeed a true reflection of parents' thoughts and manners etc. Nature seems to determine most characters, nurturing only modifies and cultivates them sometimes to the final stages of finesses and near perfection; as no one is born yet who would be perfect.

These days there are several ways of assisted reproduction available to a seeking patient. The most common practice, of course, appears to be the In Vitro fertilization or IVF, which has stood the test of time and continues to enjoy much popularity. Since the birth of the first test tube baby Louise Brown in 1978 the practice of IVF has become a universal source of pleasure for hundreds of thousands of infertile couple the world over. In fact Professor Edwards the founder of the technology has spoken about the impact of his invention on global masses on British TV many times. Of course Steptoe, Obs and Gynae specialist, who performed the miracle in Cambridge is duly recognized the founder of this technology. But both deserve the credit for the introduction of the IVF to the world. Al-Azhar approved of it in a fatwa issued on March 23, 1980. Since then many Muslim countries have adopted it as legal and moral way of reproduction through medical assistance.

The famous Al-Azhar fatawa had outlined the parameters quite clearly allowing artificial insemination with the husband of an ovum obtained from his wife, fertilization in Petri dish and transfer of the fertilized zygote into the would be mother. But only if medical reason justifies it, and must be carried out by an expert physician licensed and trained in the field of reproductive technology.

Al-Azhar clearly banned the involvement of a third party whether giving sperm, ovum, embryo or uterus. In fact any such practice would be called adultery. And Quran has given clear mandates on the punishment for adultery in Sura-e Noor. (24:1–3).

The Al-Azhar fatwa also highlighted the fact that after the death or divorce a frozen or preserved sperm from the husband can't be used for IVF. Furthermore any extra embryos obtained in the process of IVF may be employed for successive cycles in the woman married to the sperm donor husband, but only as long as they are lawfully married.

The most important part of this edict from Al- Azhar was that "all forms of surrogacy are forbidden."

Jama Al- Azhar is the leading Islamic source of education, study of fiqh and guidance for the entire Muslim world but mainly the Sunni section of Islam, Most Middle Eastern countries and major parts of Ido-Pak subcontinent as indeed Malaysia Indonesia, et are mainly Sunni Muslims and an edict from Jam Al Azhar is second only to an edit from Imame-Kaaba. But the Shia Muslims who in a significant majority in Iran, Iraq, Lebanaon, Bahrain, Azerbaijan, Baku, Balkh Bukhara and about 20 % in Pakistan tend to seek guidance from their Maraajae Taqleed either in Iraq or in Iran. Strong difference of opinion exists on main theological issues, but none disagree on the permission of IVF. Same cannot however be said about third party gamete transfer as even the Shia ulama differ between themselves on this highly sensitive issue.

The Human Fertilization and Embryology Act of 1990 (HFEA) controls the activities that fall under the banner of assisted reproduction technologies in the United Kingdom. The act defines as follows: "A woman shall not be provided with frequent services unless account has been taken of the welfare of any child who may be born as a result of the new treatment … including the need of that child for father"… [4].

No doubt the moral responsibility of both parties namely the physicians and the parents have increased tremendously with the introduction of modern technologies.

Gametic Intrfallopian Transfer (GIFT) is a very innovative procedure that is currently undergoing development. It involves harvesting the eggs from the ovary, followed by mixing them with the sperms of the donor and depositing the resultant mixture into the fallopian tubes through a catheter. The details of the technology and their intricacies are best left to the experts to elaborate, dilate, discuss, accept, or discard.

Then there is the Zygote intra fallopian transfer, where the egg is fertilized outside the body and then introduced into the fallopian tube.

In surrogacy a hired uterus is employed to sustain the pregnancy, and in cloning an asexual reproduction is carried out employing a single cell to produce genetically identical individuals. Human cloning is globally banned, but animal cloning for scientific experiments is allowed even in Islamic countries like Iran.

Different cultures have different viewpoints on the subjects as controversial as these the whole philosophy of assisted reproductive technologies has many ramifications, currents and counter currents. And a lot depends on the religious or secular approach. Besides the traditions and cultures have their own influences and affects that deeply influence such matters.

Islamic view point is determined the by the Quranic dictates and the Sunnah. Islam grants special importance to the families, the children and their rights and parents obligations it clearly admonishes any preference of boys over girls. In the pre-Islamic days, the pagan Arabs abhorred the birth of a daughter. It was considered to be a source of shame and disgrace to the father. It was therefore a common practice to bury the newborn girls alive.

In 2007, a documentary was shown on the British TV about similar practice being carried out quite blatantly in India. It was about a God-faring woman who has taken upon herself the arduous task of running an institute for the unwanted baby girls, pleading with people not to kill them, but to let her institution take care of them. It was an eye opening documentary of a modern country immersed deeply in ignorance and hypocrisy, despite its unprecedented progress in recent years. In Pakistan Maulana Edhi runs a similar institution for unwanted babies, both male and female and only God can duly reward him for his untiring efforts.

Islam, like most religions condemns baby killing. Islam has vehemently admonished mankind and particularly the Arabs, who killed their baby girls soon after birth during the *Ayam-e-Jahalia*, from such inhuman practices. Allah has given clear verdict that He alone is the provider and the sustainers, so how dare one kill a child, lest it become a liability.

Islam is a perfect code of life. It grants special significance to the institution of marriage granting the rights to the bride to sign a pre-nuptial contract that the West is now considering to employ in view of the rising number of divorce cases and their ensuing settlements etc.

Marriage, in Islam is the only institution that guarantees the propagation of joy through lawful means of producing children. Surrogacy and adoption do not enjoy much popularity and commendation. However, since infertility is a clinical condition, that brings unhappiness to the husband and wife both, the medical therapy that may be able to bring the joy of parenthood to the couple is considered ethical and moral. It is also accepted by the religious scholars, provided the whole process involves no other than the lawful husband and wife, in the process of assisted reproduction.

Since the covenant and contract of marriage is holy; and in the traditions of Prophet himself it must remain sacred and noble. Any violation of the contract witnessed by the Almighty Allah and endorsed by the families and friends cannot and must not be violated through the unethical practices.

The problem arises when the husband has severe oligospermia, and the wife is keen on having her own biological bay. Many cases of unwarranted use of sperm have been seen in the clinical practices in many countries following the code of Islamic ethics; and they are obviously breaking the sacred code.

Islam allows polygamy, but with many preconditions and not without specific reasons. One such reason is the desire to have a child. If a woman is sterile, and the husband wants to have child, then he has an option, to marry again. Divorce grants a similar right to the woman if the husband is sterile and the woman wishes to have a baby.

Until recently the concept of divorce amongst Hindus and Sikhs was an extremely unusual notion in India. It was considered sacred to be married once only. The heinous custom of Suttee was abolished by the British as they saw many an innocent woman burnt on the pyre for no fault of their own. The practice was banned only in the eighteenth century, and sometimes you still hear an odd story of the widow burning herself after the death of her husband; but that is simple cruelty. Nowadays, of course marriage can be abolished through divorce in modern Hindus and if the child bearing is an issue another marriage may be solemnized. Adoption was a standard way of continuing the lineage, particularly the *jagiradari* or fiefdom, or the dukedom in India. Britain while exploiting the weaknesses of the dying Indian empire changed the rules of inheritance in the early days of British expansion in the eighteenth century, disallowing the transfer of the dukedom to the adopted child. It gave many an Indian princess sleepless nights, finally depriving many of their descendants of the luxury that might have passed on to their adopted off springs. In fact without belittling the courage and contribution of the famous Jhansi ki Rani, it may be said that the major reason for her valour and untiring efforts to overthrow the British rule was just that. She had no male heir and the British would not allow her to adopt a child to take over the reins of her state.

If only the ways of assisted reproduction were available to those *Rajas* and *Ranis*, they would have certainly employed them to the fullest!

China is the most populous and obviously the most rapidly progressing country in the world. Before the death of communism, the old guard Mao had strictly forbidden the Chinese to have more than one child. The policy continues to remain in practice today and is arguably an obvious restriction on autonomy of the Chinese people. Most Chinese do not approve of the assisted reproduction methods, though with Hong Kong now becoming the part of mainland China, and Shanghais competing with Paris in its wealth and glory, one cannot say that the Chinese are remaining truly faithful to Maoism, and Confusciusm. In recent times IVF has been allowed to the infertile couples.

In today's Britain questions are being raised by children thus born. They want to know the exact identity of their biological father. And it is due to be allowed as the latest thought in the UK goes. And that may create many a socio economic problems. Of course the moral issues arising out of the whole process is another issue in its own right.

Africa has always suffered with poverty, ignorance, famines, and now AIDS. Like any other society, the Africans also treasure children, and the family is considered to be incomplete without them. A woman who is either barren or otherwise childless may adopt a child from the clan, the tribe or more often from within the family. Surrogate motherhood is also allowed in some societies, when a woman may arrange for her husband to impregnate another woman, with a clear understanding that the surrogate mother is in fact carrying the wife's child in her womb; only to be handed over to the wife for all times to come. A child thus born is reared with immense love of the whole extended family, indeed the clan.

Since Islam is a dominant faith in many parts of Africa, polygamy is always preferred to the later practice in most such situations.

A breakthrough occurred in 2006 for the desire of having babies through good, bad or simply innovative techniques is the concept of 'Bartering of human ova.'

Human fertilization and embryo research authority is the supreme body in England matters of embryonic research. And that obviously includes IVF and stem cell research the two most interesting, challenging and controversial topics, these days.

Recently the said authority has passed a mile stone bill. granting permission for "harvesting of human ova" for stem cell research. It means that IVF clinics can now obtain accessory ova, for research. The incentive given to women would be in the form of financial compensation. The volunteers will be granted subsidy for their IVF treatment, which otherwise costs around £3,500 or more. So according to this bill if the volunteers would allow the spare eggs to be harvested for laboratory research the authorities would grant them subsidy.

Like many other innovative researches across the world, this one have also ignites sparks of controversy. There are proponents and there are opponents of the act. A famous ethicist has strongly condemned this practice and labeled is a "bartering of the ova" rather a catchy phrase one might add.

Ethical and moral issues are always complex. They are not easy to solve or else they would not be called "ethical dilemmas."

In the present issue there are a few difficulties that are obvious to a student of medical ethics. One such major problem is that of 'organ sale' after all the volunteers will be offered money for the ova.

So the question is as to how we draw a line between good clinical practice of harnessing the latest technology for the welfare of mankind, and those that are simply unethical and downright condemnable.

No doubt the assisted reproductive technologies shall bring joy and happiness to many perspective parents, and that is the finest service that medical know how can provide. Relief from agony and provision of joy are simply the best service any doctor can provide. But if the motive is purely greed then one cannot help but condemn the greedy physician and his associates in doing so. After all the patient comes to him seeking expert help, and despite the fall in the ethical levels and demise of good old trust the physician is the one the patients puts the trust in. thus if the physician has a twisted ego and greedy, he can create a situation to suit his desires.

The technology itself cannot be blamed; it is the motive behind it and the objective in applying the technology that matters. Islam does not approve of Utilitarianism. Islam teaches us is definitely a lot closer to Deontology and that is that the means and the end should both be good.

May 23, 2012

All faiths teach their followers to bow to their creator. The word *Rab* is commonly used by many religious faiths in Indian subcontinent, such as Muslims and Sikhs, as indeed many Hindus. The word *Rab* means, the 'provider of sustenance;' i.e. the 'Sustainer.' We acknowledge with utmost humility that *Rab* has power over everything. He is omnipotent and Almighty. He has created and provided sustenance to all that moves on this earth or its belly. And all that flies in the skies and beyond, or indeed terminal plastered to the deepest layers of the darkest oceans. In the series usually produced by David Attenborough, such creatures are shown in the bottom of the oceans, that are simply too massive to even wobble let alone move. And yet, they are fed by the Almighty. Some strange phenomenon heralds in a flood of tiny fish and other creatures towards the immobile monster who simply gulps them down. Whether it's a natural current or some chemotactic scent released by the immobile monster or simply a command from the *Rab*, that the seemingly quiet and sedate ocean randomly develops these ripples that bring the food to the creature of God that stays otherwise immobile.

Rab has promised to sustain the world. In Quran there are many verses that remind us that each creature brings in its own sustenance at birth. There is no shortage of *Rizq*, except created by mankind through its evil acts.

Crops over crops of wheat, barley, and corn are burnt down in certain countries to maintain the price index of the produce. Most of the highly fertile lands in England are kept barren, so that the European food price index, net work does not get imbalanced. The farmers are paid by the EU, not to grow, a thing, in the name of maintaining the wild life, so that the prices of crops and their produce can be manipulated. One country would only produce butter in the EU, another only cattle for meat, a third one simply consume the products and so on.

In the US and surely in many other countries, live chickens hatched out of the hatcheries are simply killed by dumping them as soon as they are born, into the earth and it is a well documented fact that in the Asian countries, where the cattle is generally poor in health and mother cows produce only a limited amount of milk, the farmers destroy the male calves as soon as they are born so as to save the milk to sell in the market.

Every creature on earth, in the bottom of the ocean or in skies brings its sustenance at birth from its creator, albeit the mankind deprives many of them through its greed, avarice or simple tyranny.

The most inhumane of *Rab's* creatures in indeed the human being!

Evolution of mankind indeed its very formation under the command of the one who says '*kun*' and *Fayakun* follows is thoroughly studied by the scientists. Some of them are distinct and discernible as the embryologists inform us. The are:

In Sura- Al- Furqan (*25: 54*), it says:

"It is He who created man from water, then gave him consanguinity and affinity. Your Lord is omnipotent. Regarding the actual development of a human being, the holy Quran speaks as follows in Surah-e- Al- MuMinuun" (23: 12, 13).

"We created man from the quintessence of clay.

Thereafter we cause him to remain as a drop of sperm (nutfah) in a firm lodging (uterus).

Thereafter we fashioned the sperm into something that clings (Alaqah), which we fashioned into a chewed lump (Mudghah).

The chewed up lump is fashioned into bones which are covered with flesh.

Then we nurse him into another act of creation.

Blessed is Allah the best of Creators."

We shall later on look at the possible medical translations of Alaqah and Mudghah. But let us first read another Ayat describing the stepwise genesis of man.

"If you have any doubt, oh men,

about being raised to life again,

(remember) that we created you from dust,

then a drop of semen, (*nutfah*)

then from something that clings (*Alaqah*),

then from a chewed-like lump (*Mudghah*), complete in itself,

and yet incomplete (*ghayr mukhallaqh*) (of flesh shapes and shapeless). Which differentiates (That we may reveal the various steps to you). We keep what we please in the womb for a certain time.

Then you emerge as a child (infant),

Then reach the prime of age.

Some of you die some reach the age of dotage when they forget,

What they knew having known it once. (sura- e Al Hajj 22: 5).

At yet at another place the Quran speaks as follows:

"Oh men, fear your Lord

who created you from a single cell,

and from it created its mate,

and from the two of them dispersed men and women

(male and female) in multitudes.

So fear God in whose name you ask

Of one another (the bond of) relationships" (An-Nisa 4:1).

Embryology is a fairly established discipline. Indeed it is so established that Ian Wilmut the miracle scientist of cloning is called an embryologist, as a compliment to being a geneticist.

All physicians have to study anatomy and embryology, thus the life cycle of a fetus. There are several stages in its morphological development. At least four are clearly described in the noble Quran.

1. Fertilization of an ovum by the sperm (Zygote).
   A. Morula formation.
   B. Blastula formation, when rapid cell division takes place
   C. Embryo formation and differentiation of three germinal layers, tissues and organs.

2. The time period that lapses between conception and formation of an embryo is described clearly in the text books of embryology as 2 weeks. During this period there is a possibility that the Zygote may abort. The time period between conception and formation of a Zygote is 3 days. It is also possible that the Morula or the Blastula may abort. But more often than not the embryonic stage would pass safely and after the formation of a fetus at about 8 weeks. If all goes well, the baby will be born after 36 weeks of gestation. Between 8 weeks and birth a new addition occurs to what has been accomplished between the time of conception and differentiation of the germinal layers. So in fact the actual genesis of man stops at 8 weeks, what continues later is the growth and development.

When we examine the *Ayat* mentioned above we find exact replication of the prices in both the *Ayat* without an iota of variation in the process of genesis. Let us recap. It is first the sperm, (*nutfah*), then its safe deposit in a firm lodging (Zygote), then *Alaqah* (Morula), then *Mudghah* (Blastula), then into a form flesh shaped and unshaped bones covered with flesh, (Embryo). And then the mandate that some will emerge to live until an appointed time others may be dislodged if God so desires.

The clear definition of the steps in the development of mankind is yet another illustration of Quran as a living miracle of Allah. To a scientist it is a lot easier to define the attributes of Allah such as Al-Qadir, *Al-Khabeer, Al-Malik* and *Al-Baseer* than to a lay man. How simple it is to understand the omnipotence of God if we apply our mind to solve the dilemmas of the universe.

The genesis of mankind is quite evident from the *Ayat-e Ilahi* as described above. However the religious scholars believe that the *Surah-e Noor* describes the evolution of animal kingdom (all those that move), beginning the very life at its nascent phase in water, then the development of those who crawled (or swam), the two legged creatures, than those who moved on their four limbs. Was it the Australopithecus or its more ancient relative that the Quran refers to is highly debatable.

The *Sura-e-Baqarah* has clearly described the superiority of Adam over all creatures including the Angels, on the basis of 'Ilm' the Knowledge that was given unto him. Thus it is the author's belief that the genesis of life began in the water going through a process of evolution over an appointed time, to eventually develop into a quadruped. But the Darwinian missing link is quite evident if we combine the *Ayat* in Surah-e *al Noor* with Surah-e al *Baqarah*. Indeed there was no missing link. Mankind was clearly created from the clay into a sperm, then into an *Alaqah*, then *Mudghah* then a shapeless mass (an Embryo), then a fetus, whereas the rest of the animal kingdom underwent a physical and nonphysical evolution. That's why man is called by Almighty as *Ashraful-Makhlooqat* (the supreme creation).

The question that remains to be answered is the time at which the life actually was infused in the developing human being. Was it at the stage of Morula, etc. or could one call a sperm the progenitor of life?

There are several traditions which define the time period of 'ensoulment.'

1. "Each of you is constituted in your mother's womb for 40 days as a *Nutfah*, then it changes into an '*Alaqah*' for an equal period, then a '*Mudghah*' for another equal period, then an angel is sent by Allah to breathe a soul into you" [5].

2. When 42 nights have passed over the sperm drops, Allah sends an angel, who shapes it and makes its ears, eyes, skin, flesh and bones. He then asks the Almighty, "o Lord, is it a male or female?" The Almighty then grants what He wishes and the angel executes it [6].

The first Hadith clearly defines the time of ensoulment as 120 days from the time of fertilization. And the second *hadith* says that the differentiation of organs and systems begins after 42 nights following fertilization.

What needs to be investigated is the discrepancy between the embryologist's version of development from the day of conception through the stage of organ differentiation (embryo). The Zygote, as we have been taught, is defined as the fertilized ovum, which migrates down the Fallopian tube, staying there for 3 days, before finding its habitat in the uterus. The Morula formation is a rapid transitional stage, which instantly is replaced with a more stable phase of rapid cell division thus lending to the formation of the Blastula. This period may not take more than a week, before the Blastula converts into an embryo. The whole process from conception up to the formation of an embryo takes no more than 2 weeks. Then the gradual development and differentiation of the embryo continues for another 6 weeks, before a miniature proto type human being is formed. And this is called the 'Fetus'. The fetal stage begins at 8 weeks and continues for the next 28 weeks, before a baby is born.

If we interpret the embryologist's observation in the light of Quran and *Ahadith*, there is no discrepancy whatsoever in the actual phases of the development. However what needs to be answered in the light of the Divine guidance is the actual time period when the life (ensoulment) actually begins. The *Ahadith* are quite clear in defining the time as 120 days, from the day of conception, which would mean that, the Zygote, the Morula, the Blastula and even the Embryo up to 120, whereas science informs us it has already begun the differentiation into multiple organs and has gone on from about 15 days after conception well beyond into the fourth month of pregnancy when the ensoulment actually happens.

Thus one may argue that the life actually does not begin until ensoulment sets in; i.e. 120 days after conception. We know that it is the ensoulment which indeed differentiates life from the stage before life the so called vegetative stage of life.

Therefore the question that arises is that if Eugenics or Gene Splicing or any more advanced technological intervention is carried out at any stage before ensoulment, one may not be interfering with life at all. Is it so? We cannot answer that question forthwith. It surely needs a great deal more knowledge than we possess at this stage of discoveries. Islam encourages '*Ijtehad*' in all such matters where further research is needed.

The more interesting debate that arises in my mind is with relevance to the Law of Inheritance. The modern Genetic technology has allowed the scientist to freeze an embryo for an indefinite period of time. And to use it at a latter day by employing a surrogate uterus or indeed a lab for the end product.

On the question of the ownership of the embryo many cases of law suits have emerged in the western court. Many a battle are still to be fought on these indiscernible Waterloos. The General Medical Council of Britain debated this issue (GMC News, 5 April 2001) pointing out that the embryo actually belongs to the parent whose gametes have been employed, as follows:

"Advances in medical science raise many ethical issues – and the press has fuelled controversy with its coverage of designer babies, conjoined twins and the couple who conceived a child to be a marrow donor for its sibling".

Where should the ethical lines be drawn? Some GMC Council members have been giving GMC News their personal opinions.

Here is an interesting academic debate that took place several years ago, but carries a forward looking message for all those 'magicians' in the medical profession, who wish to exploit the scientific discoveries of the modern days. No one is allowed to play God, faith or no faith. Science means to benefit the mankind, not harm it. Eugenics may be employed for eradicating defective genes, but not for decoding them and modifying them to develop a Frankenstein. It is therefore necessary that we pay serious attention to the leaders in our profession, who speak with ample experience behind them a vision for the future. This discussion happened some time ago but is quite relevant even today, albeit a few restrictions have since been imposed on the embryonic research as indeed on the rights and obligations of the physicians.

Professor Sir Roddy MacSween, former chairman of the Academy of Medical Royal Colleges, said: "Doctors have made designer babies largely in response on public demand. It is not surprising, therefore, that parents who seek designer babies are becoming more prescriptive in terms of gender and in terms of having them for socio-therapeutic reasons such as marrow transplantation".

"I understand it is legally accepted that an embryo belongs to the couple from whom the gametes were obtained: the doctor is simply the custodian of the tissue." Sir Roddy added: "In the case of the Siamese twins, the parents views were overruled by the judiciary: the medical profession was divided in the views because it saw the issue additionally from the human/emotional aspects an out just from the calculated approach of the legal mind. It is important in such matters that the GMC – and other professional bodies – give balanced guidelines and advice and do not allow the tabloid press to sensationalize."

Sir Barry Jackson, the past president of the Royal College of Surgeons, once said "As medical science advances, related ethical issues will become ever more frequent and ever more difficult. There are rarely if ever, absolute rights and absolute wrongs. Each problem needs detailed separate debate to enable a balanced view to be achieved, while recognizing that public attitudes and culture are in a state of continuing change".

"There was a time when it would have been considered unethical to operate on the human heart!" added Sir Barry.

However, GP Dr. Surenra Kumar believes unacceptable lines are being crossed. He said: "If we try to interfere with nature too much, we'll end up with a world which is not to our liking. There was a recently reported case in Italy where somebody gave up an embryo because it wasn't a boy and all his other children were girls. That's just one of many cases where people are treating human life like commercial products. Having a baby for a medical purpose, so it can be a marrow or organ donor is absolutely horrendous. It is extraordinary that anyone would think of creating a life to be used and abused".

"In the case of the Siamese twins, I understand the parental situation and their belief that it's tantamount to killing one of the babies. But it's situation where to have to take an educated decision on whether both die or one dies, but the other has a good chance of a reasonable life."

Doctors' primary duty is always to the wellbeing of the child or potential child, Dr. Liz Bingham pointed out. Their supplementary duty is to discuss with the parents or potential parents the implications of any actions they may take. She saw two main issues about designer babies: "Trying to obtain a disease-free child when there is clear evidence of a risk of a genetically transmitted illness is now largely accepted. Where we run into problems is when people want to create a designer child for social reasons, selecting gender, hair color, eye color, intelligence and so on.

"With the Siamese twins, the principles are pretty much the same, but complicated by trying to balance the right and benefits of the two children, and involving the parents. Doctors should make sure the parents understand what the options are."

Donor conceptions are another difficult area in Dr. Bingham's view. "How can one begin to consider the interests of an individual if the prime motivation for creating that individual is someone else's benefit? You can't get consent. With a living sibling there is a chance of obtaining consent at an appropriate level".

"If a child is conceived primarily to save the life of a sibling and it fails, how is it going to affect the emotional development of the child and the emotional bondage to the parents? It risks children becoming commodities and it denies their humanity, but one may fully understand why parents might do it.

"Effectively, people are trying to design an immunologically compatible baby. That is fraught with problems and the grief of parents if the transplant fails may further complicate their relationship with the child."

Medical advances open up some highly important questions which should be addressed before the techniques are used, Dr. Bingham concluded. "It's not something only doctors can decide. Society as a whole must play a major part in deciding for itself, but doctors must inform society of the technical and scientific issues."

The decisions for geriatrician Dr. Jane Wood are of a different nature. "Our ethical questions are concerned with the quality of life, not its length," she said, "A daily issue for us is whether, if something can be done, must it be done?

6.1.3   Quality of Life

"It's a matter of how far you pursue the diagnosis and options. We don't have age restrictions, but some cardiac surgeons don't want to risk their mortality and morbidity figures looking bad by taking on lots of frail, elderly patients. But some will help these patients despite higher risk of stroke or dying. I don't have competing annuities about allotting resources between young and old, as I'm concerned entirely with looking after people of 75 and over."

Dr. Wood added: "With somebody completely demented, how far do you pursue diagnosis and treatment, when they can't give consent? In some cases, you don't even know exactly what's wrong. That's where the technology bugs are. Relatives want everything done – they ask for the patient to be resuscitated if they have a relapse, but is that humane and sensible? I spend several afternoons a week talking about these cases and some relatives are very insistent. I base my judgments on the quality of life possible and I have to be guided by GPs and families on what people's quality of life is everybody's different; that isn't one size fits all."

The laws of inheritance are described at length in the *Sura An-Nisa (4: 2–12)*. These *Ayat* describe all the rights beginning from the orphans to the free men and particularly women, at length and with great precision.

It goes without saying that Islam attaches singular importance to the rights of individuals as it equally attaches due importance to the obligations of such individuals. Islam proclaims justice, and justice is a form of equitable distribution of rights and obligations.

However such laws of inheritance only apply to living beings. As we have already discussed that life does not begin until ensoulment; i.e. 120 days from the time of fertilization. Thus one is obliged to conclude that an embryo may not possess any right of inheritance. The *Ayat* clearly mentions that some will see the day of light an reach an age of dotage, others may not be born as living beings. The Fetus on the other hand may have the rights of inheritance, but who can be sure of its live birth. After all a small portion of fetuses may have the rights of inheritance, but who can be sure of its live birth. After all a small portion of fetuses also arrive but dead.

Therefore let us summarize as follows:
1. The sanctity of life is clearly mandated in the holy Quran and described in *ahadith*.
2. The genesis of Adam is clearly mentioned in the holy Quran. The life began in the water from a single cell. Whether or not Adam went through the process of evolution is quite debatable. The Quran describes the stages of metamorphosis in *Surate Noor*, all moving life without mentioning Adam.
3. The intrauterine stages of human being are defined at least two places in Quran very clearly.
4. The clear timing of actual ensoulment does not become obvious from the Quran but is supported by *Ahadith*.
5. It appears that the human life actually does not commence until the 120 days after conception.

6. The embryo and the stages before that are therefore pre-life objects and do not possess human rights and have a vegetative life.

7. Genetic engineering for the good cause may be permissible at the pre-embryonic stage as it does not tantamount to interference with life.

8. Islam's Fundamental principles of ethics are (A) Dignity of mankind (B) Justice and (C) Benevolence. So if medical science can be harnessed to do good based upon justice and benevolence then such an act can only be called ethical. If otherwise it has to be evil, thus unethical.

9. Islam gives the final mandate *"Amr Bil Maroof Wa Nahi Anal Munkar;"* i.e. "Enjoin Good, Forbid Evil."

10. Islam is the Divine religion based upon the welfare of mankind, Promotion of *'Khair'* and prevention of *'Shar.'*

In most Western countries, particularly in the Academics and seats of higher learning, Ethics Committees have existed for many years. They determine and decide the actions carried out by the staff, whether therapeutic, or experimental, or indeed innovative and research based. All those matters which are out of the norm must be subjected to a thorough evaluation by the ethics committees. Over and above these local committees there are regional and national bodies that eventually control the matters. Such committees comprise of the scholarly people, who represent not only the Subjects, but also the faiths and cultural beliefs of the given society, for we must understand that while the principles of ethics do not change, they must take into account the cultural and social aspects of the given community into account.

Medical Practice in each country depends on the needs of the community, the level of knowledge and expertise available and the facilities that exist. In most developing countries a rat are seems to dominate the attitude and the direction of medical care. There are many countries, where even the safe drinking water is not available and yet in the same countries one may find the most advanced equipment available in abundance, mostly going to waste. Obviously someone unethical sitting at the top must have used the financial resources for his ulterior motives, neglecting the community or the national needs.

The alternate medicine, the quacks, the faith healers and even the clairvoyants are all having a gay time in many a developing country. Who is responsible for such unethical acts, and why indeed are these malpractices even allowed, let alone flourish! One wonders.

6.2 Surrogacy

Infertility is a major issue in many parts of the world. It is a bone of contention in many families in Pakistan, India, Egypt, Iraq, Iran, Nigeria, and elsewhere. Many infertile women are summarily divorced in these countries, even if the husband may have oligospermia or worse. Somehow in these countries a woman continues to be battered by the male folks. The common examples of ultimate torture of women by men folk is often seen on the media, in the villages of Pakistan. A *wadera*, or *chaudhry* or a *malik*; i.e. a duke of sorts may humiliate his subjects at his discretion. Many women are raped, tortured, and openly disgraced by their husbands or other family members at the slightest of pretexts. Therefore infertility is a major cause of humiliation in these societies. Either they are simply divorced and left to suffer all their lives as no social services available the populations in these countries.

A woman labeled as infertile may not be able to get married ever again. Stigmatization of women is a common practice in the entire developing countries. Despite their innocence such women live a miserable life. Alternatively many men folk get married again to a younger woman in the hope of having their biological child. The husband may have the joy finally, but the infertile woman ends up as a target of perpetual ridicule, both by the new bride as well as the inlaws.

Is it therefore fair to say that the world belongs to a man and woman has a secondary even tertiary position. Islam abhors injustice and such an act where a woman is discarded because of none of her faults. So what is the answer? Either a man should be educated enough to understand the situation and appreciate the problem, thus giving the woman, respect that is due to her, or a woman may be duly compensated according to Islamic law and divorced respectfully, giving her measures for sustenance and upkeep.

These and many other situations have raised question about a possible alternative of lawful surrogacy.

In Iran this issue has been discussed at many levels, Islamic republic of Iran is a very pragmatic and practical state where learned religious scholars and their highly educated and masters of their subjects physicians and surgeons discuss and debate such gray zone areas, that do not have straight forward answers. Third party gamete transfer is one such issue. (Relevant fatwas can be read on the web).

In the late 1990s *Ayatullah Ali Khama*nei issued a fatwa effectively allowing the donor technologies to be used by trained medical personnel. According to his edict the *Ayatullah* al *Khamanei* said that an ovum donation 'is not in and of itself legally forbidden', as long as the ovum donor and the infertile mother abide by the Islamic code of parenting. Accordingly the child born of a third party gamete transfer, in this case an ovum will have the legal rights to inheritance from the ovum donor. The recipient mum will be akin to an adoptive mother.

Regarding the sperm donation, *Syed Al Khamenei* edicted that a baby born of a sperm donation should follow the name of the recipient rather than donor father, but the baby can only inherit from the biological father; i.e. the sperm donor while the infertile father will be considered to be akin to an adoptive father [7].

Ijtehad is the backbone of Shia limb of Islam. No consensus has so far been reached upon this fatwa, as elites like *Maraajae Taqleed Ayatullah Ul Uzma Aga Syed Sistani and Ayatullah Said Al -Tabatabai al Hakim* advise caution against third party gamete transfer.

Many issues need to be sorted out by the Ulama. The scientists have achieved what was an impossible task not long ago, but no medical technology can be applied without strict parameters of morality and ethics or you may end up with a Dr. Frankenstein. Islam does not permit human experimentation without control and legislation. The main issues in this context are whether sperm donation should be allowed at all. Perhaps not. One is reminded of a jurisprudential challenge that the Caliph Omar was approached to resolve. A group of women raised a voice that polyandry should be allowed, as Islam believed in equal rights to all its followers, Omar turned to Ali As for help who is the greatest jurist of all times. Ali invited all women who came to protest to bring a small pitcher of water on their next court appearance. He then asked them to pour the water in a large pitcher. They all did so as advised. Then Ali asked each one of them to seek and withdraw her part of the contribution of water. Obviously they were all puzzled and could not comply. Ali As then gave his verdict that just like the pitcher here, a woman's uterus is like a pitcher, from which a mixture of fluids cannot be isolated; hence leading to the time honored question of lineage and heritage. No wonder Caliph Omar Al- Khatab once said that "let there not come a day, Ya Allah, that Omar be faced with a dilemma and Abul Hasan may not be there to resolve it!"

Another question that crops up is that of identity of the donors. Is the third party gamete transfer permissible if the donors are anonymous? In the recently published 'Time Magazine' article that is the very problem that has been identified as a major issue too. Anonymity may breed incest. If the donors are known, however, it may lead to the dilemmas mentioned in the 'Time' article. In the Jewish community in Israel third party gamete transfer mainly of the sperm is allowed only if the identity of the donor is known, though the rabbis prefer a non-Jewish sperm donor to avoid future genetic incest.

Islam is a masculine religion and the genealogy is traced through the fathers rather than mothers, even though mothers are given a prime place in the family structure. As Prophet said that "the paradise sits at the feet of your mother" obviously advising us to serve our mothers as an act of *Ibadat*. We also know that each person would be called to task on the *Roze- Qiyaamat* by his or her mother's name and yet we all trace our heritage through our fathers. In fact it was part of Indo-Muslim culture in the days gone by, when a family attendant would stand on a podium and recite the whole genealogical chart verbatim, and by heart to the entire audience, tracing it through fathers and grandfathers right up to Ali as, if the groom happened to be a Syed. It was practiced to ensure the purity of the blood line. This man Friday was called the Mir Kalas. He is an unknown entity in the present Indo Pakistani Muslims. According to some cynics, Mir Kalas, dropped off the Kalas bit and became Mir Sahib as soon as he crossed the Khokhrapar border in 1947. Mushtaq Yousefi the legendary satirist of the subcontinent has described this practice in writing that "in order to glorify the past of his ancestors to justify his current success and grandeur of sorts this character hastily replaced the cut throat knife in his ancestor's hand with a scimitar!"

Both ovum as well as sperm donation do not require the physical aspect of the practice of *Mutaa* as no actual physical intercourse is necessary for the task. Frankly speaking, a sperm donation may lead to a thousand complex outcomes. Most *fuqaha* have in fact banned it and are rethinking on the whole concept of third part gamete transfer and surrogacy.

It is also worth a note that, special clinics have been set up by some assisted reproductive specialists; i.e. Gynae/Obs clinicians in Lebanon, who happily display the fatwa of *Ayatullah Khamanei* and other *Marajaes* to provide third party gamete transfer services to women of a certain age. Many non Shia women from the neighboring Arab countries travel down to these clinics to have their biological children, not for their maternal joy which is of course the prime mover instinct, but also to save their marriage and their dignified life per se.

A few years ago, a documentary was made by two Asian legends namely, Zia Mohiulddin and Shabana Azmi, which portrayed the plight of women in the subcontinent who would travel any length to have a baby. Regrettably the story in this documentary was true albeit pathetic to say the least. It just displayed the activities that go on in certain so called saintly abodes where infertile mums gather to reap the harvest of their supplications. It was indeed a shocking story which truly displayed the plight of such women and their exploitation at the hands of the so called godly men or women. It was good enough evidence to supplement the argument that something ethical, legal and moral must be available to such desperate women.

Mutaa was a lawful practice in the times of the Prophet SA and also in the caliphate of Ali AS, and Hasan AS. It was banned by Caliph Omar Bin Khatab for an unknown reason. There are very specific limitations and conditions that must be met for a *Mutaa* to be accomplished. It is not an uncommon practice in Iran even today. *Ayatullah Rafsanjani* had realized that after the Iran Iraq war imposed by Saddam the tyrant on Muslims of the region, thousands of women were widowed and had no way of supporting themselves. Some could get married the second time but not all. Fears of corruption dominated the society and the biological as well as financial needs of the women folk also mattered. Therefore *Mutaa* was legalized, for such widowed women. Later on due to the biological needs of having one's own child this practice was extended to such couples under strict moral and religious control. No physical intercourse involving the third party was allowed; albeit a verbal contract was made between the ovum donor woman and the infertile couple, mainly the husband. Once the ovum under the *Mutaa* contact was obtained the *Mutaa* contract was dissolved, and the gamete employed to fertilize with the husband's sperm in a Petri dish, finally implanting into the infertile woman's uterus through IVF.

Sunni ulama have their strong views about *Mutaa* and its legal status; hence they do not approve of this practice. But that is mainly based upon the condemnation of *Mutaa* by the second caliph in the early days of Islam. Islam is a religion of progress and Quran is the last living testament. No further revelation will ever come now. Shia faith has a strong source of guidance available to them through the lineage of Prophet As. The twelfth Imam continues to provide guidance to them from his occult abode through the *Maraajae*, according to their belief. *Ijtehad* is the fundamental difference between Shia and Sunni Islam. While the later continue to practice Ijtehad through the learned scholars in *Hauza-e Ilmias* of *Najaf al-Ashraf* and *Qumm*, the Sunni Islam stopped any form of Ijtehad long ago. They instead opted for *Qias*; i.e. opinion of the celebrated few. It may be argued that such a *qiyas* may be faulty, heavily biased and often inappropriate as many stakeholders may be involved in a critical decision making.

No religious argument may be complete unless it is either proved in the light of *Quranic* teaching or that of Sunnah. But Shia have an enormous advantage as their

Ayama lived for about 250 years after the Prophet. Therefore they had enough time to study their life and practice of the Prophet through the direct descendants, rather than relying on the hearsay.

An interesting article was published in the 'Time Magazine' [8]. It was an eye opener for a student of ethics. It described the global impact of frozen sperm that America is exporting. Just like a thousand other items, USA seems to have the largest share in the export of sperm load. American sperm is not only being employed by the ethnic groups in USA but also being exported to many countries across the western world. The sperm bank is continuously being replenished by the young and handsome men of many ethnic backgrounds, allegedly for the benefit of matching women. The major bulk of clients are infertile couples or women of a certain age who cannot conceive naturally. According to a report from the WHO in 2006, between 60 and 80 million infertile couples existed in the world. Apart from the infertile couples there are many homosexuals who are also employing these banks.

The ethical and legal dilemmas are also discussed in this article. Since anonymity for the sperm donor is an option in the west, many legal and ethical issues are bound to arise with the passage of time. One such dilemma is that when the children born out of such an association grow up, they may wish to seek and identify the paternal source of their birth. Therefore the legal rights of inheritance and rights of sharing the estate of the father/donor crop up and the matter becomes even more complex as the sharing of names etc. may also come to the forefront. And that is not the end of the issues anyway; sometimes incest may ensue out of such an exercise. In fact many of these sperm donors have sold their load for huge amounts of money to sire innumerable children in Australia, New Zealand and England, etc. Fortunately not in the Muslim countries, where it is banned to import the sperm load from USA or elsewhere.

The reason behind the discussion is that the matter of third party gamete transfer is currently being discussed in the Islamic world. No definite decision has been made on a universal basis as the Sunni *fiqh* rejects and condemns the practice altogether. Shia *mujtahids* are currently debating the issue in Iran and elsewhere.

A group of doctors in Liverpool happened to discuss the issue of *wudu* at an after dinner meeting. The question was of course of washing the hands and upper limbs up to *Marafiq* or from *Marafiq* to the fingers as also the question of washing the feet. In *Surae* Al -*Mayida* (5::6) Quran has clearly mandated to wash the face and the hands up to al *Marafiq*; i.e. elbows and perform *Massah* of the head and the feet up to the ankles. There is no mentioning of washing the feet head or neck as is practiced by our brethren in Islam. The word used in *Surate* al-*Maieda* with reference to face is '*Agsulu;*' i.e. washing and for the feet, the word is '*amsahu;*' *i.e.* touch; i.e. '*masah*' not wash. The argument was made that the Prophet did so. Most were Sunni Arabs and Pakistanis, but one of them was a Shia, who for obvious reasons became a target of ridicule. Eventually the Shia doctor asked the audience, whether they knew the habits and habitat of their fathers and their families better or the next door neighbor. Obviously the answer was in favor of the family. Therefore the Shia colleague argued that we practice what our fathers did and so forth until we see what Imam Ali did who saw what Prophet did, as he was the closest disciple of the prophet besides being the father of his progeny. The argument was thus resolved, but of course as the Quran teaches us. "*La Ikraha fi deen.*" So do as you please as long as you and I worship Allah ST and call Prophet SA the last messenger, and believe that the noble Quran is the final divine testament.

Another important question that troubles the ethicists is that of abortion. Islam condemns and prohibits abortion except under very special circumstances where the life of the mother may be at risk, and that too after the Ulama have consented to it. In current situation of modern technology, many a new dilemmas have cropped up. One such situation is that of pre-genetic diagnosis (PGD).

This technology has allowed an early diagnosis of a fetal abnormality through amniocentesis, sometimes as early as 14 weeks after conception. So if a baby is discovered to have a significant developmental abnormality such as Spina Bifida, Signs of Chorea and Athetosis or a Trisomy of sorts, then should the mother be allowed an abortion or not. Many a diehard would label it illegal and unethical as they believe; it is Allah's will to put some of us under great stress and test our patience. Well, Allah is *Rahman* and *Rahim*. He has promised not to put any form or quantity of stress more than one can bear upon His humble subjects. It is the Mullah who wants to turn Allah the beneficent to Allah the Curser'. One cannot condone such an argument. One just have to look at the lifelong plight and total commitment of such parents who happen to have a deformed or disabled child, to appreciate the justification of allowing an abortion after the PGD has established a possible Trisomy or something similar.

The burden of disability will not remain confined to the family alone, as we often see in an egalitarian society like Britain where the state provides for the care of such a child throughout his life. Islam is a religion of shared commitments and responsibilities. The concept of *Maslaha* is applicable to such situations, where the loss of an unborn disabled child may save the family as well as the society of a huge economic, social and moral burden. Many such children, if left under the care of the institutions in the West do not have any quality of life. Their misery is often caught on camera by some clandestine cameramen and shown to politicize the issue from time to time. In the developing countries where no social services of any sort area available, except for the NGOs and God-fearing folks like Edhi sahib, the plight of such children remains indescribable.

In Iran as indeed in Saudi Arabia, PGD is now formally allowed and the mother may then opt for an abortion before the completion of 14 weeks, to save herself and her family as indeed the society of a burden of disability and untold misery.

Only last week a mother of a disabled baby told me that she was informed at 20 weeks of gestation that her baby was going to be abnormal. She was given the option of an abortion, but she declined. Now, she said, she regretted it as she could not bear to see the child's suffering as he was deaf, blind, mentally retarded and a Trisomy of sorts. She literally cried and hoped for a mercy from the Lord, that some way of easing his and her family's agony. She in fact said that one day when the baby suffocated and turned blue the family did not do much; but he came back and their agony continues. She was a non-Muslim, and quite educated, believing that her emotional blunder was proving to be life time of misery for the child and a huge burden on the family and states resources. In fact she even mentioned that the baby had no quality of life, would it not be better if he passed away!

Life is not without its share of joys and sorrows. They both go *pari-passu* almost like the hand in glove. Islam means reform and happiness of its followers. It does not want to see its followers suffer in any way or form, either financially, morally socially or politically etc. The whole philosophy of Islam revolves around the pivot of goodness against evil of sorts. Any technology that is developed to benefit the mankind has to be ethical and morally right albeit it should be duly measured in the light of *Quranic* teachings and the practice of Prophet and his *Ahlal-beit*.

Feb. 27, 2013.

While travelling to Barbados to the American University of Barbados, medical school, I read an item in a British newspaper supplied by the crew on the plane, which can only be described as a hair-raising story in a moral world.

A woman had a genetic cause of her infertility, so she and her husband decided to go for surrogacy. The family came in to help share the burden. At a fertility clinic, the sperm of the husband and the ovum of the wife's sister was fertilized in vitro, and implanted in the uterus of the sister of the husband. The fruitful pregnancy resulted in the birth of a pretty girl, who obviously was shown as the apple of not one but eight pairs of eyes. Everyone was happy with the outcome.

Could it be called an incest? I wonder. Or a practical solution to a family ordeal? I leave it to the reader to decide.

March 26, 2013.

A medical seminar was held in Kufa, Iraq in March 2013. Many interesting lectures, papers and presentations were made by international speakers. They were all very informative and enlightening, but one presentation deserves a special mention here. It was a paper presented on the subject of surrogacy. It was not only hugely informative but also resulted in an extraordinary debate, as Fatwas after Fatwas were discussed by the learned audience in favor or against a certain injunction. In the city which hosts several *Marajaehs*, of no less a standing than *Ayatullah ulUzma Agha Syed Sistani and Agha Bashir Najafi*, an intellectual debate of such immense proportion was not just refreshing but instructional for all of us who live in other countries and are not directly exposed to the scholarship of the *Marajeh*.

Excerpts from that invaluable presentation is reproduced here with the permission of the author, Professor Azhar Mousa Al-Toriahi of University of Kufa Medical School, a Gynaecologist and Obstetrician of repute.

She began by defining surrogacy and described its various types.

1. Traditional surrogacy

    It involves artificial insemination of a surrogate mother with the sperm of the intended father by IVF, IUI, or home insemination. Thus the newborn will be genetically related to the father as well as to the surrogate mother.

2. Traditional surrogacy and donor sperm

    A surrogate mother is artificially inseminated with donor sperm via IUI, IVF, or home insemination. The child thus born is genetically related to the sperm donor and the surrogate mother.

3. Gestational surrogacy

It is employed when the intended mother is unable to carry a baby to term due to such factors as hysterectomy, diabetes, malignancy, etc. In this technique her ovum and intended father's sperm are transferred into and carried by the surrogate mother. The resulting offspring is genetically related to its parents but the surrogate mother has no genetic relationship.

4. Gestational surrogacy and egg donation

If there is no intended mother or the intended mother is unable to reduce the ova, the surrogate mother carries the embryo developed from a donor egg that has been fertilized by sperm from the intended father. By this method, the newborn is genetically related to the intended father and the surrogate mother has no genetic relationship.

5. Gestational surrogacy and donor sperm

It involves a process when the intended father is unable to produce the sperm, thus the surrogate mother carries an embryo developed from the intended mothers egg (who is unable to carry a pregnancy herself) and donor sperm. Thus the newborn is genetically related to the intended mother and the surrogate mother has no genetic relations.

6. Gestational surrogacy and donor embryo

Then she went on to define the people who may choose a surrogacy. She mentioned heterosexual with one partner being infertile, a woman unable to carry a pregnancy or go through a pregnancy, single men wanting a child, or the gay male couples, women who may have defective ovarian function, a congenital anomaly or agenesis of uterus, or recurrent loss of pregnancy or indeed failure of IVF implantation.

When the intended parents are unable to produce either the sperm, the ova, or an embryo, the surrogate mother can carry a donated embryo (often from other couples who have completed IVF cycle with left over embryos). The child thus born is neither genetically related to the intended parents, nor indeed to the surrogate mother.

Dr. Mousa then took us through an invaluable historical journey, beginning with an antique law called the 'Law of Antiquity' which allowed another woman to bear a child for a couple with the male half of the couple as the genetic father (A very learned scholar in a private discussion later on pointed out the Biblical story of Abraham, whose wife Sarah could not conceive thus asking her maid Hajer to bear a child for her!). I have no comment to make, but according to the law of Antiquity, was that indeed a surrogacy.

According to Professor Mousa, the Babylonian law, indeed originating from the very lands where this presentation was made, allowed this practice and infertile woman could use it to avoid a divorce.

Many social and cultural variations, led to the paving of the pathways for surrogacy over the last hundred or so years.

In 1870, it gained popularity in China, for the infertile couples to adopt a male off spring to become a legal heir to the family treasures. This practice, one has to say is quite different to the earlier Islamic practice, where a male child could be adopted but did not gain the name nor the inheritance of the father as for example in the case of Mohmmad bin Abi Bakar or Zaid Bin Haris. But parallels can be drawn here in terms of the practice of adoption *per se*.

In 1930s, oestrogen became available in the US, which made many hormone induced pregnancies possible.

In 1944, like in many other fields, Harvard took the lead when Professor John Rock became the first person to fertilize human ova outside the uterus, and in 1953, many researchers elsewhere successfully performed the first ever cryopreservation of the sperm.

The journey forward continued and in 1971, the first Test Tube baby was born in England, through a process called IVF. Louise Brown lives a normal life.

The first ever law in favor of surrogacy was written by a Michigan lawyer Noel Keane as a surrogacy contract. And in 1985, a woman carried the first ever successful gestational surrogate pregnancy.

The battle upon the egalitarian and hereditary rights of the baby born out of surrogacy has been a controversial topic in all circles, both in the Western secular society as indeed in the Islamic culture.

In 1986 a surrogate child called baby M was born in the US. A conflict broke out between the surrogate and biological mother on the custody of the baby as the biological mother refused to give the custody to the couple with whom she had made a surrogacy contract. The court of law in New Jersey found that the biological mother was indeed the legal mother declaring the surrogacy contract illegal and invalid. But in the best interest of the newborn the custody was granted to the biological father and his wife instead of the surrogate mother.

So the dilemma is not that easy to resolve. It continues to baffle the law makers of the secular and religious schools even today.

In 1990, another legal battle in California ruled that the true mother is the one who intends to create and raise a child.

Then in 1994 Latin American fertility specialists convened in Chile to discuss that assisted reproduction and its legal and ethical status.

On the other hand the Chinese government banned gestational surrogacy due to the legal complications of defining true parenthood and possible refusal by surrogates to relinquish a baby.

There is a logical connection according to this learned speaker, in the whole process of surrogacy. It could be:

1. Partial or genetic contracted motherhood, alias traditional or straight surrogacy, in which the gestational mother is impregnated with the sperm of the commissioning father (usually through artificial insemination). In such cases the gestational mother is both genetic and gestational mother of the child; however she relinquishes her role of being asocial mother to the commissioning mother.

2. Complete or gestational contracted motherhood (also known as 'Hostor gestational surrogacy'). Employing the IVF technology, the intended parents produce an embryo that can then be transplanted into the surrogate mother for her to gestate and give birth. In this process the pregnant woman makes no genetic contribution to the child; however she is (nevertheless) the child's birth mother.

In some cases, particularly of infertility, this may be combined with the use of donor sperm, or donor eggs, in creating the embryo for transfer.

There are huge controversies in the battle of surrogacy that continue to rage. Many arguments have been put forward against surrogacy, including the consideration of the interests of the surrogate mother and the rights of the baby.

Other arguments include such questions as the speaker pointed out:

1. What if the surrogate mother or the commissioning couple change their mind (during the term of pregnancy)?
2. What if the surrogate pregnancy ends in a miscarriage or indeed multiple births?
3. What if the newborn has a serious disability? Who should then bear the responsibility of lifelong care, expenses, etc? In a welfare society such as the UK, where the social services exist and provide excellent support, but what about the developing nations, where the family alone has to bear the burden of disability? And what if the resources are inadequate, insufficient or outright non-existent? In the matter of disability the American surrogacy centres strongly advise that any issues regarding potential birth defects and abortion should be discussed and resolved before signing any contractual surrogacy agreement.
4. And what are the legal and hereditary rights of the newborn?
5. Finally, should a large payment (commercial cost) be made for surrogacy, or is altruism (minor expenses such as travel and lodging, etc.) enough?

Many ethical dilemmas were presented by the learned speaker, such as the factor of exploitation, bullying, coercion, threat or bribe, etc. to the surrogate mother, particularly if the customers or clients are wealthy, resourceful and powerful, targeting a weaker, vulnerable woman to be their surrogate.

Then there is the question of human rights. Can a woman make contract for using her body parts to the benefit of a third party? What if during the pregnancy, the surrogate mother decides to have an abortion or indeed refuse an abortion, if the baby is discovered to be abnormal (say through PGD)?

The million dollar question is the definition of motherhood. What indeed is the relationship between genetic motherhood, gestational motherhood, and social motherhood? Is it possible to socially or legally conceive of multiple modes of motherhood and/or the recognition of multiple types of motherhood?

Then there is a strong question about the rights of the baby to know of the identity of any/all of the people involved in that child's conception and birth.

In order to avoid incest, the Jewish authorities have made it mandatory that the identity of the sperm donor be recorded and known to the IVF team, since a huge bulk of sperm load is being shipped out of America, where there is a fairly large portion of Jewish population, who could be the potential donors. Professor in her excellent discourse went on to discuss the legal issues involved in surrogacy.

For instance, the surrogacy agreements are enforceable, void, or indeed prohibited. Does it make a difference whether or not the surrogate mother is paid commercially or takes an altruistic approach and is simply reimbursed for expenses?

Another question is that of an alternative to post birth adoption for recognition of the intended parents as the legal parents, either pre or post birth.

She opined that, despite the fact that the law differs from land to land, some generalizations such as, the historical legal assumption, that the woman giving birth to a child is that child's legal mother, and the only way for another woman to be recognized as the mother is through adoption (usually requiring the birth mother's formal abandonment of parental rights).

Are there any psychological implications upon the parents and the newborn out of surrogacy? Well there certainly are, but matters are currently under evaluation by the research scientists. On the face of it, it appears that surrogate mothers rarely had a difficulty in relinquishing the rights on a surrogate child, and that the intended mothers showed greater warmth (motherhood) than the surrogate. But that is certainly open to debate. Motherhood is a natural bondage. One cannot imagine a mother give a birth and not have any love for the bay she carried in her womb for nine months feeding her with her own blood!

On the other hand researchers found no difference in negativity or maternal positivity or in fact child adjustment in a study involving 32 surrogacies according to this speaker.

Many surrogate children, however are beginning to raise questions about their lineage in the contemporary western world.

During the discussion many participants believed that the third party gamete transfer is not allowed in Shia *Fiqh* as in other fiqahas. Grand ayatollahs *Ali Khamenei* was often quoted as an authority, who is pragmatic and believes in active research, allowing surrogacy as long as a 'no touch no see policy' was adopted in third party gamete. No one however, objected to the IVF or the surrogacy if the process involved only the married couple. The audience did not speak much in favour of *Mutta* or *Sigheh* as it is not practiced by the Iraqi Shias as it does in Irani Shias.

The debate heated up as the matter of life and ensoulment became a topic of discussion. Several Ayat were quoted by learned participants including the Al-*Haj* and Al *Muminoon*. Dean Saeed, a learned speaker went on to quote *Sura Al -Tariq* (86:6,7) "He (man) is created from a drop emitted: proceeding from between the backbone and the ribs(loin)." Dean Saeed said that it referred to the descent of testis. The embryologists may enlighten us on that point.

During the luncheon someone raised the question of disposal of aborted fetuses or indeed amputated parts of Muslims in the Western countries. Frankly I had no answer, but one scholar categorically insisted that such parts must be buried according to the Islamic rituals. But does that happen? I am not sure. To the best of my knowledge, all such items are labeled as 'clinical waste' which is disposed of observing specific rules and regulations, and duly incinerated. Yet another dilemma for the competent authorities to solve.

References

1. Afzaal A (1994) Do we have to accept the Westss solutions. Daily Dawn Magazine, 23 Sept
2. Quran 23:12–14
3. Rahman F (1987) Health and medicine in Islamic tradition (change and identity). The Crossword Publishing Co, New York, pp 114–115

4. Human Fertilization & Embryology act 1990 (HFETA). Updated in 2008. www.legislation. gov.uk/1990/37
5. Sahi al Muslim. Kitab al Qadr 5:496. www.sunnah.com/Muslim
6. Sahi al Muslim. Kitab al Qadr 5:499–500. www.sunnah.com/Muslim
7. Inhorn M (2006) Making Muslim babies. IVF and Gamete donation in Sunni versus Shia Islam. Cult Med Psychiatry 30(4):427–450
8. Time magazine (2012) 179(15):34–35

Organ Transplantation

No doubt, organ transplantation is a marvel of modern-day medicine. The first human organ to be transplanted from a cadaver was the cornea in 1905. It was soon followed by blood transfusion, which occurred in 1918. The first successful kidney transplant happened in 1954. Following the melodramatic first-ever cardiac transplant surgery in 1967 by Prof. Christian Bernard, a plethora of varied forms of organ transplant surgeries have come to stay. The world literature is inundated with the moral and ethical implications of this art of giving someone a new lease on life. A whole new world of terminology and nomenclature has developed.

Much debate and discussion goes on globally on the scarcity of organs and the growing demand for them. Questions are being raised about ethical and moral ways to procure these organs.

In an article published about the scarcity of donation of cadaver organs, Caplan and Viring [1] ask a question: 'Is altruism enough?' They say that the lack of an adequate supply of cadaver organs and tissues for transplantation to those who need one poses a major problem to the transplant community and to those responsible for national policies. Generally, in the USA, a combination of altruism and voluntarism is practiced to obtain an adequate supply of donor tissues. A rapid rise in the field of transplant surgery, better survival rates, and a new lease on life to an otherwise crippled patient have increased the demand of donor tissues manifold. The ongoing shortage of organs and tissues in recent years, according to these authors, calls for abandoning the law of 'altruism or voluntarism' in favour of either a 'market system' or a 'system of presumed consent'.

These researchers say that a survey of the impact of federal and state laws requiring that requests for an organ donation be made to the next of kin when a death occurs in a hospital show that attending medical personnel do not enforce these laws with sufficient zeal.

Some years ago, a story of an immigrant was published in a British newspaper, which said that a healthy man was smuggled into England and his healthy kidney was removed and transplanted into a paid donor at a private hospital. The British government had to intervene and ban such practices. Many patients from the UK go as medical tourists to the Indian subcontinent for a transplant, to buy a kidney for a paltry sum of £200–300. In many cities and rural areas on the Indian subcontinent, many poor are robbed of their life savings by unethical, dishonest doctors, all in the name of 'saving a life'. Of course, the majority of such commercial transplants result in rejection, not to ignore the financial loss and total lack of morality.

Similarly, in India and many African countries, parents have been reported to sell their organs to buy a suitable organ for transplant in one of their wards, akin to the practice of blood donation, where a relative may donate blood to the bank as a substitute for a matching pint for his relative. That is why some people argue that payment for an organ may be pragmatic and not a condemnable practice after all, provided it is fair and square. Iran has taken the initiative in this regard: the authorities are looking at possible payment for an organ, not as a price for an organ, which as we know is priceless, but more as a token of gratitude.

In the late 1980s, organ transplantation took off, and many centres in the world became totally dedicated to these practices, sometimes with unhappy results and often unethical practices, particularly in developing nations. The highest health authorities, therefore, took up the task of laying down rules and regulations. The guiding principles of human organ transplantation advocated by the World Health Organization a few years ago are discussed here. It was indeed a pioneering document of huge historical importance. Many modifications have since come into force and are now available on the Internet.

7.1   Guiding Principle 1

Organs may be removed from the bodies of deceased persons for the purpose of transplantation if:
(a) Any consents required by law are obtained, and
(b) There is no reason to believe that the deceased person objected to such removal, in the absence of any formal consent given during the person's lifetime.

This principle highlights the importance of consent for allowing the removal of organs after death. It does not specify about a living will or an advanced directive, which are essential and must be available before an organ harvesting team commences its action. Besides the second component is akin to the present-day presumed consent, which is still under discussion in the British parliament in 2013. The impression that one gathers from this original guiding principle is that authorities were already thinking along the lines of presumed consent or a standing order in the form of a will or of 'opting out.' So if such a written instruction is unavailable, the organs may be harvested. For a Muslim patient, this principle may not be acceptable. Also, for many Asian communities, the family may not agree to such an action and will disallow the removal of organs through a 'vali', or surrogate representative of the deceased. That is why I strongly believe that every Muslim patient, as all others living in the Western world, must write a will when they have their full senses to leave clear instructions as to how the body or its parts may be disposed.

7.2   Guiding Principle 2

Physicians determining that the death of a potential donor has occurred should not be directly involved in organ removal for the donor and subsequent transplantation procedures, or be responsible for the care of potential recipients of such organs.

Definition of death remains controversial. As early as the 1990s, systems had not yet developed and people were still groping in the darkness. Nowadays, ethics committees have an extremely important role to play. Each hospital or institution providing health care has an ethics committee. Controversial issues are referred to it, and they make decisions keeping the medical as well as ethical principles in view. No doubt the final decision regarding death is now made by a team of physicians, including the primary care giver, a neurologist, an anaesthesiologist etc. The process of determining the final brain death is described in an essay [2].

In most countries, there are organ donor registries maintained by a central authority, and in the event of a tragedy where the potential organ harvesting may be applicable an urgent warning to the transplant teams is sent, and the harvesting team, which is always more than those involved in the rivalry care and death decision-making team, arrive to harvest the organs while the transplant surgeons are preparing the recipient, so no time is wasted in the process. Special flying squads work with extreme efficiency in the UK and USA. In many developing countries, things need to improve further.

7.3   Guiding Principle 3

Organs for transplantation should be removed preferably from the bodies of deceased persons. However, adult living persons may donate organs, but in general such donors should be genetically related to the recipients. Exceptions may be made in the case of transplantation of bone marrow and other acceptable regenerative tissues.

This guiding principle is almost universal. Living donors are always preferred, and many stories of amazing degrees of altruism ate published in the dailies in the UK. It is also a preferred practice in many developing countries. Regrettably, though, despite the formal teaching of altruism and sacrifice as the essential component of many faiths, lack of knowledge and fear of unknown do not make this practice of living relative donations such a common practice. Until a few years ago, many developing nations of the Muslim faith would not allow a cadaver donation, considering it to be akin to a desecration of the body, forbidden in Islam. However, due to massive campaign by the teams of transplant surgeons and active support of the government, and later on the *Ulema*, it has now become a norm, and many cadaver donations are now available in countries like Iran and Pakistan.

An organ may be removed from the body of an adult living donor for the purpose of transplantation if the donor gives a free and informed consent. The donor should be free of any undue influence, coercion and pressure and sufficiently informed to be able to understand and weigh the risks, benefits and consequences of consent.

7.4   Guiding Principle 4

No organ should be removed from the body of a living minor for the purpose of transplantation. Exceptions may be made under national law in the case of regenerative tissues.

This principle is still a challenge, but is being maintained *perse*. Minors have a special place in ethical terms and the responsibility for their decision rests upon the parents, or the guardians. It is easy to understand in many Asian communities where the family bondage is strong and the children are a treasure. But as we witness in the Western countries, the very fabric of the society has broken down. Many teenage mothers give away their newborns to the caretakers and foster homes. No doubt they are above board and look after the adopted children as their own. But have we not read horrible stories of child abuse, torture, tyranny, inhumanity even murder of children by irresponsible adults. As I write this paragraph, I have just read in a local daily, that a young mother killed her three young babies, then committed suicide, for an unknown reason. Only last week a horrible monster was sentenced for life, when he was found guilty of putting his own house on fire to kill five of his children, in collaboration with his wife. What monsters!

Even the church is not spared. Horrible stories of the pious bishops and cardinals keep surfacing from time to time, about the sexual abuse of young boys in the holiest of places. Not much different are the stories of religious schools or *madressahas* in Afghanistan and the tribal belt of the northern Pakistan. A few years back a woman Sikh playwright exposed the sexual abuse of young girls in Gurdwaras, in a play staged at a theatre. Perhaps only one or two shows could be staged, as the Gurdwaras chiefs, along with a fairly large Sikh community arrived at the theatre, smashed the windows and stopped the play forthwith. The poor woman who wrote the play had to hide for the fear of her life. Such is the might of the religious zealots!

So when you see such situations, you are bound to think of the children as extremely vulnerable human beings, who must be safe guarded against such monsters including the use of their parts.

The media is inundated with the news of the abuse of young girls by a very famous TV celebrity, who abused his authority and power to destroy so many lives. He was even knighted! So when you read such stories and the damage they have done to the society, you are obliged to agree with the authorities that the rules must be even stricter than they are at presented. No fish, big or small may be allowed out to escape the net, as these animals destroy the very fabric of humanity.

It is therefore mandatory in the UK that any one involved in any form of contact with children, be it educational medical or leisure, must obtain a clearance from the police through a Criminal Record Beareu (CRB) form, before taking up a job.

7.5   Guiding Principle 5

The human body and its parts cannot be the subject of commercial transactions. Accordingly, giving or receiving payment (including any other compensation or reward) for organs should be prohibited.

This guiding principle is ageless. In fact, one of the reasons it was promulgated was that in early days of transplantation surgery, organ trading was the only way one could obtain an organ. The poor and the needy folks were, obviously, exploited and and their bodies abused. Even today, organ trading is a major problem in the developing nations of Asia and Africa. Medical tourism has grown into a major industry in India and Dubai. Mega hospitals and rich investors have opened up many centres in these and other parts of the world where people, who can afford it, travel from Europe and USA to obtain a kidney, or other bodily parts at a commercial cost. I personally know a particularly unethical surgeon who did this trade for many years in Karachi, and was eventually caught and thrown out of the country, or one may say, escaped the country before he could be arrested.

Iran has yet again shown a very pragmatic approach, as discussed elsewhere in this book. Since organ trading could not be stopped despite the lawful breakdown of such a trade in the black market, the government, through a NGO, pays a certain amount of money to the donor, which is officially recorded. It is in lieu of service to humanity, in order to save a life, and basically takes care of the donor and his family for a defined period of time, both in health and sickness.

Many other nations, which are struggling for the organs, are currently engaged in debates and discussions amongst the various stake holders to look into the nuances of this practice.

7.6   Guiding Principle 6

Advertising the need for or availability of organs, with a view to offering or seeking payment, should be prohibited.

Commercialisation of of organs in any form or way is an abominable act. Allah has created the man in His image. The human body as indeed the soul belongs to Allah and it has to return to Him. Some scholars define it as a 'loan' from the Creator. So if the body does not belong to you, then how can you advertise, sell or commercialise some parts of your body? It is akin to a 'khayanat' in an 'amanat'. That act is therefore to be condemned. No faith or philosophy condones dishonesty, theft or pilferage from a property that does not belong to you.

7.7   Guiding Principle 7

It should be prohibited for physicians and other health professionals to engage in organ transplantation procedures if they have reason to believe that the organs concerned have been the subject of commercial transactions.

Medical profession is a noble profession. The oath that we take at our graduation specifically mentions that we will not bring any form of dishonour to our profession. Trust and confidentiality are the founding pillars of this profession. Honesty and integrity are the building blocks, and conscience determines our acts on daily basis. So it goes without saying that if a physician suspects that an unethical act has taken place either in organ procurement or its supply he must not use his expertise in transplanting such an organ despite the fact that it may save a life. By doing so, he not only he betrays his oath but also burdens the soul of the recipient.

7.8   Guiding Principle 8

It should be prohibited for any person or facility involved in organ transplantation procedures to receive any payment that exceeds justifiable fee for the services rendered.

This guiding rule is basically a corollary of the rule discussed earlier. Commercial activity of any nature involves two parties in a transaction. This rule advises against the receipt of money for transplant. Usually it is the poor who are either forced by circumstances or by force to accept the money. And usually it is the rich and the powerful who make them an offer that they can't refuse. Obviously it is wrong, but as mentioned before legal and lawful payment to the needy donors may not be far from today to become a routine globally, as is happening in Iran.

7.9   Guiding Principle 9

In the light of the principles of distributive justice and equity, donated organs should be made available to patients on the basis of medical need and not on the basis of financial or other considerations.

Justice is the most salient of all factors that build the structure of ethics. This topic has been discussed in a full essay in this book. Suffice it to say here, that distributive justice must form and remain the guiding principle in not just organ transplantation but all other services, too. In a social welfare state like the UK, health services are provided based on the principles of uniform and equivocal distribution of services, for anyone who needs it rather than who can afford it. But in many countries, including many advanced nations, many poor people have either no access to health services or only to limited services. So if an organ is available, the poor person may not get the benefit of it as he is usually at the bottom of the list. It is seen daily in Indo-Pakistan, the Middle Eastern and many African and Far Eastern countries. Those who can't afford it do not get it. And that is extremely sad. In some countries there are charitable organisations support such patients and they must be applauded for such services. But they are not everywhere.

The Islamic viewpoint was elegantly expressed by the Islamic Organisation for Medical Sciences, Dr. Hassan Hathout, a famous Muslim ethicist wrote in 'Topics in Islamic Medicine' [3].

"The individual patient is the collective responsibility of society, which has to ensure his health needs by any means, inflicting no harm on others. This comprises the donation of body fluids or organs such as providing blood transfusion to a bleeding person or a kidney transplant to a patient with bilateral irreparable renal damage. This is another *firdh*-e- *kifaya* – a duty that donors fulfill on behalf of society.

Apart from the technical procedure, the onus of public education falls on the medical profession, which should also draw the procedural, organisational and technical regulations and the policy or priorities."

Every one agrees that the organ donation should never be the outcome of compulsion, family embarrassment, social or other pressure or exploitation of financial need. Furthermore, a donation should not entail the exposure of the donor to harm.

The medical profession bears the greatest burden of responsibility for laying down the laws, rules and regulations for organ donation during life or after death by a statement in the donor's will or the consent of the family, as well as the establishment of tissue and organ banks for tissues amenable to storage. Cooperation with similar banks abroad should be established on the basis of reciprocal aid.

In early days of Islam it was decreed that if a man living in a locality died of hunger being unable of self-sustenance, then the community should pay his money ransom (*fidiah*) as if they had killed him. The similitude of people dying because of lack of blood transfusion or a donated kidney is very close. Two traditions of the Prophet seem to be quite relevant in this respect. The one is: The faithful in their mutual love and compassion are like the body... if one member complains of an ailment all other members will rally in response [4]. The other tradition says: The faithful to one another are like the blocks in a whole building ... They fortify one another [4].

Allah describes the Faithful in the Quran saying: "They give priority over themselves even though they are needy" [5]. This is even a step further than donating a kidney, for the donor can dispense with one kidney and live normally with the other ... as routinely ascertained medically prior to donation.

If the living are able to donate, then the dead are even more so: and no harm will afflict the cadaver if heart, kidneys, eyes or arteries are taken to be put to good use in a living person. This is indeed a charity ... and directly fulfils God's words: "And who-so-ever saves a human life, it is as though he has saved all mankind" [6].

A word of caution, however, is necessary. Donation should be voluntary and by free will, otherwise the dictators and the ruling tyrants in many countries will confiscate people's organs, thus violating two basic human rights: the right of freedom and the right of ownership.

In the society of the Faithful, donation should be in generous supply and should be the fruit of faith and love of God and His subjects. Other societies should not beat us to this noble goal.

In fact, the Holy Book has aptly highlighted the authority of Allah in the matters of life and death, at many places. "It is He who created death and life" [7] Mankind is only a custodian of this gift of God. Hence we *say 'Ina Lillah wa Ina Ilehey Rajaoon'* [8].

Professor Magdi yakoob is a legend. At the famous Harefield Hospital in London he began his career nearly five decades ago. Beginning, as we all do, as humble junior doctor, he had vision, but not much insight into what future held for him.

However, he soon discovered that his passion lay in heart surgery. In due time he would create his own niche, and let the world following his footsteps. He pioneered the art and science of heart transplant surgery in the UK.

The late 1960s were very exciting times. Earlier on the Beatles had just returned from their whirlwind tours and the swinging '60s were in full glory. A South African Surgeon named Christian Bernard had just performed a miraculous surgery and man had landed on the moon. Of course, the overall implications on the world would be far more due to the accomplishment of Armstrong than that of the late doctor. His contribution to the art of surgery that he performed at the Grutt Schorr Hospital would earn him considerable repute, but also raise some serious ethical questions. Was the transplanted heart tissue matched with the recipient? Was the transplanted heart rejected, as the recipient barely survived a few weeks? That breakthrough surgery certainly opened the flood gates of transplantation surgery.

Magdi Yakoob has also braved many a storm in his life. Though retired for some time, the good doctor continues to benefit mankind through his charitable works, mainly in developing countries. At one stage, he was bitterly criticised for his efforts at popularising heart transplants in children.

These are not the only legends as there area many more like De Bekey and Cooley, etc. Pakistan has quite a few heroes of its own. Mulana Abdul Sattar Edhi is one. But my focus to attention in this essay is in the topic of organ transplantation. In that, one name shines above all, that of Adeeb ul Hassan Rizvi.

Like all great human beings he is also humble, modest almost shy and extremely clever. He also has the looks to match his persona. But that is not what I intend to write about, as I know he is careless about the praise as he is free of any taboos reading disapproval by many who may not like him. I wish to write about the work he and his dedicated team have accomplished in the field of organ transplantation in Pakistan.

Organ transplantation is now almost a routine practice in many countries, both developing and the developed. The panorama is changing every day. Each country has its own programme, its own resources, and its own dilemmas. Pakistan has many problems. One of them is common to most nations, i.e., the organ shortage.

Chronic renal failure is a major problem in the developing countries. Of course it is a big problem in the West also, but there area several major differences that we shall discover as we proceed in this debate.

Pakistan has a huge population of 180 million people with resources limited to suffice only a small fraction of the population. The burden of disease caused by renal diseases alone far exceeds any other country in the region. Unfortunately, the economy is struggling to survive, the resources are limited and burden of disease is perpetually rising.

Rationing of health resources is a major issue in the UK and USA [9]. Long and never-ending debates continue to echo in the long and dark corridors of the academia as indeed in the silken shinny halls of the parliaments. Managed Health care is a big issue and constantly receiving support from diehard supports of autonomy. And yet not many of them know how the other half lives.

While the UK and USA continue to be rich and famous the poverty stricken Africa and Asia barely survives on bread and water, which is often enough not quite easy to obtain.

Every one in the Western halls of ethics and morality speaks of justice. It is a fine slogan to draw attention of the ordinary mortals, but where is justice, and how do you receive it, is matter for all to guess.

Rationing of health care and distributive justice are two dominant issues in health care [10]. We in the subcontinent are quite used to rationing. So it does not sound strange to our ears when some illustrious health pundit talks of rationing the resources in the light of limited resources. For instance, Hasceptin, a life-saving drug in breast cancer, is rationed out. Some would be eligible to get it through the NHS, while others have either no access at all or have to buy it privately. It is a major debate under the caption of rationing of the resources. I have to say that the best book I have read on the subject of cancer is called the 'Emperor of all maladies.' It is written by a renowned oncologist in the USA, and because of its quality, in investigative journalism and the information it provides in a simple narrative style, it was justly awarded the Pulitzer Prize a few years ago. It highlights many interesting problems in cancer treatment.

The NHS is struggling to meet the rising cost of its customers. There are already plans in application and in the pipeline to ration out some services and make them less accessible to the patients, some of which are quite logical, but controversial. For instance, the old fashioned Royal Infirmaries of the Second World War are now being closed down. There were far too many of these small hospitals across Britain. They were good enough to serve the population for nearly five decades, but no more.

The technology has changed, the patients have changed, the disease pattern has changed and the economic conditions have changed. So the government has made clear choice in favour of upgrading the entire health care system in the UK. The trusts are being given more autonomy through the foundation programmes. Devolution of power and resources is the order of the day, and each trust has to compete with the peers to harness more clientele, hence more money. The minutest of procedure these days can bring in some cash to the trust. The KMR form has to be filled in by each doctor for as minute a procedure as an aural toilet. Tesco, the giant supermarket, has a motto, '"Every little helps." It aptly applies to the NHS of 2008.

The new mega-hospitals on the American models are overtaking the old fashioned larger and outdated hospitals. More sophistication means less access to many services for many patients. And that is a form of rationing, which has raised quite a few storms in the UK. The market economy has pushed out the corner shops that used to sell the bread and butter easily accessible to all and sundry, and the supermarkets and hypermarkets have taken over the task. These super stores will provide a wider range of services, but the waiting lists would be longer and services rationed out. Therefore, the consumer will be pushed out to buy a private health care. The concept of private health care is relatively new to Britain, though elsewhere in the world it is a norm.

Rationing in the field of organ transplantation is a standard procedure in most countries. Of course the priories are determined on the basis of seriousness of the condition rather than financial status. But not always. And that is where the debate on the sale and purchase of the organ focuses.

It is a universal fact there is more demand than supply of hearts, kidneys, livers and lungs, etc. And there shall always remain the shortage as the technology improves and progressively more patients survive the complications ensuing tissue transplantation. That has led to the reason that the scientists are concentrating on the generation of tissues to transplant through stem cell research and genetic engineering.

In the TV debate that I mentioned earlier in the essay, there were two groups of people, the patients who had suffered with life threatening illnesses prior to an organ replacement surgery, and supported the idea of organ trading and the moralists who simply abhorred the idea of buying an organ for money. One moralist went on to paint a picture of a mother in a poor country selling her kidney to buy food for her children. A couple of years ago, a news item was published in a British newspaper reporting that a man actually sold his kidney in Bombay to buy food and clothing for his family.

About a decade ago, I was visiting Australia to participate in a global meeting on Ethics. I came across a news item in Sydney, which I recall as I write this chapter. The news was quite appalling to say the least. It was a barbarous act of inhumanity. A Turkish worker was invited to a party and made absolutely drunk. He was then escorted by a pretty girl to a hotel. The next day he was discovered lying unconscious in the bath tub, filled with ice, with a gash along his lumbar region which was hastily and roughly sutured, left to fend for himself. The man was taken to the hospital by the police only to discover that one of his kidneys was removed. Apparently many such gangs operated in several countries where greed and avarice would overtake any human values.

The question is that if such is the demand of the organs and there is a growing need, what should be done? Well, the answer has puzzled all the savants and the pundits and the problem continues to baffle everyone.

No doubt, for a kidney transplant, a family member is the best option to secure an organ. In many countries, including Pakistan, it is now a norm. However, the battle that has been launched by Adeeb and his illustrious team may solve some of the issues, including the question of transplantation of organs other than kidneys.

Cadaver donation is a norm in many developed countries. In the UK, the debate goes in favour of autonomy of the next of kin allowing such a donation or withholding it, as against following the pattern seen in some EU member countries. In some of these nations in Europe, the principle of 'presumed consent' is followed. It means that the state has the authority to harvest the organs of a dead person unless the next of kin stops such an act.

In the USA, many green card holders well remember signing a card of organ donation in the event of death. Now, unless the next of kin stops the authorities from doing so, the person who had signed the consent willingly at the time of becoming a permanent resident alias alien, is liable to have his organs removed by the competent authority in the case of death.

Cadaver donations is perhaps the best answer in countries such as Pakistan. Trauma and terrorism are currently the two major killers in Pakistan and many other countries going through these difficult times. If only the public could be educated that a life of a dear one lost to trauma or an accident, may not be lost in vain, if only the next of kin could donate the organs of the victim to many who would benefit with the transplant. Obviously such grateful members of the society would offer many prayers in the favour of the donor's soul.

Muslims, believe that the soul shall return to the Creator. It was on loan. The body was nothing more than a casket, a pot, vessel to contain the 'amanat' from Almighty Allah. Now that the amanat has returned to its rightful owner, the souls which has been described in Quran as 'Amre Rabbi,' has departed and returned to Allah, the earthenware shall return to its origin, i.e., clay. What better way to become eternal than to give away the organs to kindle the lamp of not one but several lives, before it all changes into clay?

The legislation passed by the parliament in Pakistan, through untiring efforts of all those who were waging a battle to save many from dying due to organ shortage. The parliament passed the bill in 2011 making illegal trade of organ transplants a crime and encouraging close relatives or people known the recipients to donate organs provided such donors are above 18 years of age. More significantly, the legislation has been approved by the legislative assembly to harvest cadaver organs for transplantation. This wall-breaking parliamentary act will go down in the history of Pakistan as an extremely laudable piece of legislation, worthy of all the praise that can be showered upon those who promoted the cause and those who approved it. It shall also be harbinger for all other Muslim countries to adopt a similar stance and save many lives through a charitable, philanthropic and praiseworthy act, surely better than many an ibadaat.

Altruism is a common practice in most cultures. It is however more often seen in the developing nations, where a mother would often starve only to feed her baby. A father would walk miles to till the farm and sell its produce to feed the family even though he may be hungry himself. Every father in the developing nation works at the oddest of hours to make an extra amount of money, not for his personal comfort, nor to save for his retirement (as that concept is alien to the eastern culture), but to send his children to the finest of schools. Such are the examples of the parents in India and Pakistan. Such are also the situations where the dishonest traders exploit a mother to sell her body or father his kidney to feed his family. What can you say about such people except curse them?

One glorious example of ultimate altruism, voluntarism and ultimate example of service to mankind is that of a professor of Pathology. The gentleman, a learned professor knew and fully understood the human values. I knew him personally and always admired, like many others, his humility, modesty and quiet and sedate nature. He died recently, donated his organs to many dying patients, who can now live for many more years. No doubt by giving this gift of life, a mortal

Professor Memon has become immortal. The rest of us should learn from his example and sign an organ donation form, before the call comes.

One may argue that, yes, the bodily organs are being sold and purchased, and yes, there are some genuine, and possibly very compelling, reasons for people to do that, but that does not mean that this practice should be condoned. In fact, it should be condemned and stopped altogether. Legalising cannabis in Britain on the pretext that its easy availability would kill the black market has added to the problem not solved it. Many people have stated growing cannabis in their garden sheds, even in the houses rented for accommodation, but discovered by the police to grow cannabis in the bedrooms, and elsewhere.

Time has come that rationality should overtake emotional decision making in matters of organ donation, After all the fundamental difference between mankind and the rest of the biological creatures is that of the faculty of reasoning.

Imamia Medical International is emerging to be a global platform for like-minded scientists, physicians and Ulama to discuss common issues and serve mankind through the expertise available in its rank and file. In April 2012, the fifth international meeting was held in Najaf- Al- Ashraf [11]. It was a huge gathering by any standard. The detailed report is available on the IMI web page. In a debate on medical ethics, which was attended by scholars from many nations, several interesting dilemmas were presented to the panel of experts by the audience. One such question was whether an organ obtained from a non-Muslim can be transplanted in a Muslim patient in a country such as the UK or USA.

The learned Maulanas did not approve. But the ethics expert on the panel, a surgeon by profession, disagreed with the Maulana and posed a question to the audience: If a Muslim needs the blood transfusion should he be given blood only from a Muslim donor? Or should the blood donated by a Muslim be reserved only for a Muslim patient? In a pluralistic society it may be impractical if not outright impossible, unless the Sharia councils in the Western countries decides so and make it mandatory akin to the *Halal* meat practice.

Blood is infact the liveliest of all organs and a true life saver. The common practice in the West is that if a patient requires blood, unless he is of Jehovah's Witness faith, transfusion is given only after the scientific cross matching and the usual protocol of screening for HIV or hepatitis, but never checked for the race, religion, colour or creed of the donor. So, logically speaking, if the blood can be transfused then a kidney or liver can also be transplanted. Further discussion at this meeting did not pursue, but another place a question was raised as to whether a Muslim can donate his organ to a non-Muslim or not. Here, the argument is that the body being an *amanat* of Allah ST, and soul being His gift, the organ, if donated to a non-Muslim may suffer exposure to *Haram* food and drinks, etc., thus, it may compromise the sanctity of the Muslim organ and the soul. Most Shia jurists have no objection to this pratice. If a Muslim has died and his organ can save another human life then it may be allowed; after all, the Quran has clearly mandated that 'if you save one life as if you have saved the entire mankind [12].' But such are the matters where physicians and ulama should sit together, and reach a decision to guide the physicians practicing in the West.

The subject of organ donation would remain incomplete without the discussion on the topic of the timing of the harvesting. And that in turn means a discussion on the subject of death and its various definitions.

Nowadays, strict ethical principles and active monitoring have made such practices impossible to conceive, but stories continue to circulate that in China, the moment a person is executed his organs are removed and traded.

Therefore, the timing of organ harvesting is an important issue. The demand for organs is growing every day in the world, and the technology and immune-suppressive drugs are now so effective that the replacement of organs has now become a routine. The waiting list, nevertheless, continues to mount. Therefore, organ harvesting has become an essential component of the whole phenomenon. Timing of harvesting an organ is extremely important because sensitive tissues like the cardiac muscle have a very restricted time frame during which it can survive in frozen condition. Likewise, if the organs are not harvested within a specified, limited amount of time, they would become worthless. It is therefore a common practice now that once it has been determined that a person is brain dead, his organs are kept artificially alive until an appropriate time for the harvesting to be completed.

Islam is against any form of desecration of the dead body. But after great deal of debates and discussions, *fatawa* are now available in the Islamic world to remove the organs for transplantation with dignity and respect and under strict observations of the rules laid down by the competent authority.

In Iran, Imam Khomeni not only brought about a political revolution but also encouraged and promoted huge scientific advancement. When asked a religious question called *istifa*, about harvesting the organs from the braindead, he deliberated and consulted his medical advisors before issuing an edict in its favour.

Iran and Pakistan lead the Islamic world in the field of organ transplantation. The detailed Fatwa issued by Imam Khomeini is available on the net. (Akrami et al. gives in-depth analysis of the situation in Iran [13].

Iran has also introduced a policy of paying a small sum of money to the voluntary donor through a government supported NGO. It is not a price for an organ but a form of help. Many other countries are currently studying this option.

Pakistan on the other hand has performed a ground breaking task. She has succeeded in gaining approval of the legislative assembly in favour of the cadaveric organ harvesting. The legislation was promulgated recently, and should pave way for other Islamic countries to follow suit. That is huge breakthrough, for which many stalwarts but particularly Professor Adeeb Rizvi must be given all the credit for the untiring efforts, spread over a span of four decades.

References

1. Caplan AL, Viring B (1990) Is altruism enough? Crit Care Clin 6(4):1007–1018
2. www.medicalethics.imamiamedics.org
3. Hatout H. www.islamicmedicine.org
4. Sahi Bukhari, Sahi Muslim, Ibne Hanbal, by Al Noman Ibne Bashir

5. Quran, Al Hashr :9
6. Quran, Al Maida :32
7. Quran, AlMulk 67:2
8. Quran, Al Baqarah 2:156
9. Mensel PT (1992) Some ethical costs of rationing. Law Med Health Care 20:57–66
10. Jeccker MS (1989) Should we ration health care? J Ed I-Luman 10:81–89
11. www.imamiamedics.org
12. Quran 5:32
13. Akrami SM, Osati Z, Zahedi F et al (2004) Brain death: recent ethical and religious consider-
    ations in Iran. Transplant Proc 36:2883–2887

Managed Health Care

8

8.1   Rights and Obligations

Judaism, Christianity and Islam – all three of which are Abrahamic, monotheistic religions – firmly advocate the fundamental rights of human beings, rights of confidentiality, doing good and avoiding evil, and performing duties based on the voice of one's conscience. Here, however, we shall mostly concentrate on the Islamic viewpoint in all bioethical matters. Islam is the final religion. It sums up all that was revealed in other religions. Islam strongly believes in human dignity, the sanctity of life, and benevolence and justice. We shall examine these facts in the light of religious teachings.

Autonomy, beneficence, and justice form the pillars of medical ethics in the contemporary world. These cardinal rules have been described by the Muslim scholars as:

Takreem
Birr
Adl
Ehsaan

Autonomy is the fundamental principle of medical ethics. It basically means that every human being is born free and possesses the right to live as he wishes. An individual is not an abstract thing. He is a 'person' who has certain duties to perform and matching obligations to fulfil. These rights and obligations are described in many places in the noble *Quran*. Singular importance is accorded to the weakest in society, namely orphans and women (Al-Nisa: 75), and those who are well-off are given the due responsibility of providing care, assistance, love and affection to them. In other words, while the rich are bestowed with bounties, they have matching responsibilities to fulfill, not just to other individuals but to society in general and to the downtrodden in particular.

Everyone, irrespective of his position in a given society, is entitled and duly empowered to make his own decisions. "Do whatever you may wish" (Fussalit, 40), albeit not without structure and caveats. It goes without saying that the firm boundaries of what is allowed (*halal*) and what is forbidden (*haram*) are clearly defined at many places in the noble *Quran*. Thus, although Islam grants an individual rights of

S.H. Zaidi, *Ethics in Medicine*,
DOI 10.1007/978-3-319-01044-1_8, © Springer International Publishing Switzerland 2014

freedom to do what he wishes, it clearly defines the parameters of such freedom. Autonomy without any checks and balances inevitably leads to an erosion of authority, disrespect to elders, a breakdown of families, degeneration of family values and a loss of morality. Its illustration can be seen in any Western culture.

Trust and confidentiality form part and parcel of the principle of autonomy. They are of paramount importance in health care. The finest illustration of which was the singular importance of the Prophet's own physician, who was a Jew. Obviously the Prophet must have confided in him on many occasions. That also speaks volumes for this physician, who never betrayed the principle of 'Confidentiality' an ultimate virtue that a physician must possess.

The relevance of authority vis a vis autonomy of a person is further highlighted in this verse of *Quran*: "You are not the one to compel them (Qaf: 45)." Therefore, while some are granted authority over others by providence, they are instantly bound to withhold their authority and not compromise the value of human dignity.

But Islam being a religion of welfare and equity does not confine its commandments only to grant autonomy or curtail the powers for abuse of authority. It also grants equal importance to responsibility. One cannot have rights without obligations. Thus it says in the holy book: "Regarding the taking of responsibility and accountability, [beware that] every soul is responsible for its deeds" (Al-Tur: 21), and again: "Every soul is responsible for its own deeds" (Al-Mudatthir: 38).

Allah has defined the acts and their relevant responsibilities quite clearly, as the Quran dictates, "Hearing, eyesight and mind: all of those, he shall be accountable for" (Al-Israa: 36).

Thus ,at several places in the Holy Quran, mandates have been given assigning duties, as well as their rights, for individuals and the community. A balance must exist between the rights and obligations. One cannot have rights without fulfilling one's obligations.

Islam attaches singular importance to the role of the community and grants special favour to those who serve the interests of community, over and above their own interests, even personal loss. Every where the Quran speaks of individuals "duties such as Al Salat, instantly followed by Al- Zakat", i.e. an obligation to serve the needy in the community.

Autonomy and the concept of rights and obligations are therefore quite evident from these verses in the Quran. One cannot argue the importance of free will that Allah grants to an individual, in both observing the code of conduct in terms of claiming his rights as indeed in terms of fulfilling one's obligations. The essential message of the holy Quran is that a privileged person must share his blessings with those in need. We must also remember that Islam strongly urges the principle of sacrifice. It not only urges, infact grants it huge importance as we know through the yearly commemoration of the Hajj and the essential component of this religious ritual in the shape of the slaughtering of a sheep, lamb or a goat, to remember the great sacrifice that Abraham (PBUH) made at Allah's command.

There is a clear message here also, that while a Muslim's conduct must show a balance between individual's rights and his obligation, he must not forsake these smaller virtues for a more glorious cause such as the 'sacrifice.' When translated in terms of medical ethics it can have several ramifications, such as the welfare of the society at large compared with individual gains (*fard e kifayah*), or the care of the terminally ill, when even the next of kin may have forsaken him. Or indeed raising funds for charitable cause at the cost of an individual's own time, money, efforts even self esteem. Islamic teaching are so universal and the principles described in Quran so broad that one can harness them in all matters of life and draw the conclusions for guidance. The noble book is quite clear in granting due importance of individual's rights as discussed above, but it attaches equal importance, indeed more to the rights of community being superior to that of an individual. The *Quran* says:

> Who ever quickens a human being, it shall be as if he has quickened all mankind (Al-May'dah: 32).

It does not simply mean the physical only but both mental and physical awareness and awakening in all matters of day to day life. In Islam more importance is given to the rights of community than to an individual. For instance offering daily prayers, mandatory for all Muslims, are preferable if offered in the community mosque. Likewise the Friday congregation is made obligatory to be offered in the regional mosque. The annual Eid prayers are preferred to be offered in the main city mosque, the *Jama Masjid* as compared to the local mosque. And the ultimate '*ibadat*', i.e. *Hajj* is a congregation of the entire Muslim *Ummah*, where love, care, sacrifice and communal activities supersede individual acts of worship.

In Medina, which was the first Islamic state, Prophet (PBUH) demonstrated the example of communal activity by employing himself along with other Muslims as a laborer in building the Masjide Nabavi. In Medina, the city of Holy Prophet, all citizens enjoyed equal rights. Even the worst of his enemies be it Munafiqeen or the Zimmis, had rights of freedom of thought, action and access to justice. The Hadith says "A Muslim cares for his brother and protects him. And he does not fail or forsake him (in the hour of need) (Al Bukahri, Muslim, Abu Daood, Al Tirmizi, Ibne Hanbal)."

There are numerous examples in Islam, where irrespective of one's faith, justice was granted purely on the basis of judicial inquires and the proof of evidence. Thus suffice it to say that the Islamic code of Ethics grants respective cognizance to the concept of autonomy, (best translated into Urdu and Persian as Khud Mukhtari), in all matter of life.

In Suarh-e- Al Nahal (verse 49) Allah commands: "InallahYamir, Bil Adle Wal Ahsaan" meaning that "Allah commands Justice and Beneficence."

And then in Sura-e- Al Rahman (verse 60): "Should the reward for goodness be aught else but goodness." "*Hal Jaza al Ehsaan, Illal Ehsaan.*"

The most ultimate attribute of Allah is Unity, '*Ahad*' or '*Tauheed.*' This is followed by Al *Adil*, or 'The Just,' Allah is Adil, i.e. distributor of Justice. He is the final Judge of all matters of life and death.

By Justice one means decisions based upon the concept of equity. And fair play. Justice cannot be dispensed unless one knows all the facts. And in order to dispense justice one must have authority and that cannot come without deep knowledge. In depth knowledge is only obtainable through deep thinking and prudent perception, which in turn is derived from firm faith. It is this ultimate attribute called '*Yaqeen*,' that has been described as the prime mover of all just decisions by Ali Ibne Abi Talib in one of his glorious sermons in Nahjal-Balagha.

In medical ethics when we interpret the philosophy of beneficence and justice we can find very practical illustrations of them both.

In day to day clinical practice, one must remember that rights and obligations are equally applicable to the physician as well as the patient. So while the patient is within his rights to expect a beneficent approach by the physician in his treatment, it is also equally true that the physician has rights also to decline treating a patient under certain circumstances and transfer the care amicably to a colleague. This does happen in life, as sometimes a patient may be totally irrational, demanding, abusive, even physically threatening to the physician. In such a situation if the conflict can be resolved by staff, and mangers etc., then the matters may continue, but once the physician has felt abused, he becomes vulnerable to future threats and is within his rights to withdraw from the care of this patient.

I recall one afternoon, when I was running my clinic in a hospital in the West Midlands, that a commotion broke out in the adjacent consulting room, we all rushed to check it out and were appalled to see a big, burly patient shouting and yelling over a frightened physician, who was about to be hit with a fist. Obviously the security was called in and the patient was taken away. The physician in question was so scared for his life that he could barely speak.

Some years ago, it happened to one of my closest associates in Pakistan. He was not just threatened physically after a patient suffering with malignancy did not live long after the very basic intervention, i.e. biopsy. The entire blame was put around the neck of this surgeon that the use of his knife, as brief as it was, triggered off the death.

The harassment went on for couple of years, eventually forcing this surgeon to migrate.

So while the patient has a right to receive the best treatment, a physician is under no obligation to provide it. Similarly, the patient has equal rights to request change of the caring physician if he likes or indeed withdraw from the treatment for his own reasons. Here I should like to quote the illustration of Jacqueline Kennedy, who stopped all forms of treatment when she was informed of her terminal illness. She however withdrew from the world scene to spend the rest of her life in privacy, away from the eyes of the world.

One of the daily problems faced by us, who are in active practice in the NHS, UK is the pressure of patients demanding all forms of investigations, even when the physician does not consider them necessary and totally uncalled for. Many immigrants and asylum seekers are particularly worth a mention. Some of them have entered the UK from countries, where they did not have access to even the basic health care let alone advanced surgeries or sensible investigations, but they not only demand, in fact create unpleasant scenes pressuring the doctors beyond measures.

European countries have many nations where health services are either limited, or expensive, such as the artificial appliances or surgical operations. Many of these Europeans travel to the UK on the basis of a European passport, and avail of free and fine health services, appliances etc. and return to their countries. One colleague is currently conducting an audit of all those patients who presented for surgery from neighboring countries and did not return for a follow up. They were thus declared as DNA; i.e., *Did Not Attend* and discharged from the care.

An unhappy result of surgery on such patients can be quite troublesome for the surgeon as they know only thing; i.e. the doctor did it wrong. Fortunately unlike

Iraq where we recently heard horrible stories of physicians in jail or under life threat, British doctors have legal and professional protection through defence unions and medical protection agencies. The latest debacle of the Mid Stafford Hospitals has opened new avenues, which are quite worrying as the courts may now be involved in prosecuting those doctors, nurses or paramedics found guilty of neglectful death of these patients. Obviously, such apprehensions may lead to increasingly more defensive medicine, just as was narrated by a renowned surgeon to us, in Iraq.

A balance between rights and obligations, for both parties namely the physician, now called the service provider ad the patient alias client, must be maintained and respected. Many dilemmas can thus be avoided or resolved with peace and mutual understanding.

8.1.1 Rationing: Services Within Limited Resources

The first brick of the foundation is indeed the principle of justice. All other matters rest upon this foremost principle. The supremacy of truth' is upheld by all religions and philosophies. And there cannot be truthful without morality. And the most tangible attribute of morality is justice. The unprecedented advancement of technology has resulted in the cropping up of dilemmas like euthanasia, surrogate motherhood, frozen embryos, abortion, eugenics and of course cloning; some of which are discussed in this book.

In a world of falling resources and financial constraints, the new concept of 'Services within restricted resources' is developing the world over. Let us critically examine it in the context of the health care in the developing world.

Rationing of medical care means a fair and equitable distribution of services to individuals as well as the community at large. It obviously means a just and equivocal availability of medical care to all those who may need it. Rationing is not an unfamiliar word. Rationing of the flour or sugar and sometimes Petrol are all too familiar happenings in the context of many parts of the world. Some in fact imposed due to political ulterior motives of the mighty nations against countries like Iran.

Some forms of rationing in medical care are a norm, such as the provision of prosthesis or the dental implants, cardiac valves, or the hip replacement prosthesis etc. but in explicit terms the concept of medical rationing does not refer to such accepted norms. It simplicity refers to services, which are otherwise considered to be a matter of individual's basic needs in day to day care. In many countries the insurance companies control the destiny of medical care as indeed of the provider and the beneficiary, i.e. all parties concerned, have status of the strange terminology for playing a game of hide and seek, non-discrimination and satisfactory monitoring capability [1].

In a country like Pakistan which in 2013 is falling apart due to atrocities on the minorities, and failed economic state as well as governance, an ideal health delivery system can only be called 'Utopian.' Currently, the nation is allocated a paltry sum of less than 1 % of the GNP for education and about the same for its health needs. Just to make the sharp contrast in identifying the national priorities, one may cite the example of Japan, where since 1961, the nation is progressively expanding its health coverage, so that in 1997, 100 % of Japan is covered by the socialised health insurance system. The nationwide medical cost is paid 57 % by the health insurance, 31 % by the public money, 12 % by the patients and the remaining 12 % by other sources [2].

Here are the latest figures published in the Sunday magazine of Daily Dawn, Karachi, on the anniversary of the poet-philosopher Allama Iqbal, the man who dreamt of Pakistan. It is simply pathetic to say the least, for a population of 180 million, and brings neither honour nor glory to the name of Allama; perhaps quite the contrary.

In numbers

Hospitals in Pakistan 972
Dispensaries...................................4,872
Basic health units..............................5,374
Maternal & child health centres.............408
Doctors.......................................149, 201
Dentists...10,958
Nurses..76,244
Lady health workers.........................95,000
Midwives..1100
Hospital beds................................108,277

Ratios:
1,206 persons per doctor
16,426 persons per dentist
1,665 persons per hospital bed

Between 2011 and 2012:
30 BHUs and seven rural health centres have been constructed;

35 BHUs and 15 health centres have been up graded.

(Source: Economic Survey of Pakistan, 2011 – 2012)

Despite all the difficulties that UK is facing these days, no one is denied the basic treatment. The founding principle of the NHS that the care is provided on the needs and not the ability to afford it, still holds true. Perpetual complaints against the services have pushed the, and government on its back feet, though one has to say that they are not such great supporters of free health services, and would rather privatise it. But, the common man is struggling to make both ends meet due to falling national revenues, bankruptcies and the lot, but no government can dare demolish the NHS or it will be doomed.

In fact, the forthcoming so called Obamacare is a variation of the national health service. It is popular with the poor and the needy but vehemently opposed by the rich and the powerful, including the doctors.

In the world of shrinking resources rationing has become inevitable even in as boisterous economies as Japan. But, in all these developed countries, parameters for rationing are sufficiently described. It is universal truth that, the ethical basis of rationing in the field of health care has to be the covenant that binds the physician to the patient. Though unwritten, this agreement of faith is grounded in the fidelity of the doctor/patient relationship founded on the basis of mutual trust.

Thus, one can hardly underrate the importance of faith and all these attributes in the life of an individual, and that too who is so fondly described in our literature as "*Messiha.*" It is incumbent upon a physician that he remains honour bound to serve his patient within the precincts of the covenant that he signs as he takes an oath.

A patient coming from a distant hinterland of the remote areas in the developing countries in Asia and Africa may not know the various degrees or diplomas that the physicians happily advertise on their hospital boards. He is thus more often than not misguided even maltreated by a maleficent doctor. The lowly qualified is often the one who advertises his adages the most. But then who has not heard of a plethora of alternate treatment, now available even in the UK. Infact right opposite the oldest postgraduate and the most celebrated medical centre in Karachi, multiple warnings painted on the walls as graffiti reminded this author daily that deviated nasal septum could be corrected with a sniffing powder. And that the tonsils should never be removed as our home grown Tabeeb had a medical panacea up his sleeve. But, then in a society where all is ill and nothing works within the lawful parameters, one is depressed rather than feel sorry for things.

Illustrious Mushtaq Yousefi Saheb, the great humorist and satirist describes this practice in his gleeful style. He says to this effect that "one is not surprised to note that in our country a gastric ulcer is treated with '*moong ki daal,*' and jaundice with a false talisman, one feels angry that the patients do indeed recover with such anecdotes."

Informed consent is an extremely important entity in the tool-kit of medical ethics. Providing Informed consent is a process in itself. It consists of several facts of patients' care. It involves consideration of goals, alternatives, risks, advantages, disadvantages, merits or demerits, and sequelae of treatment and non-treatment etc.

Adequate consent demands clear and precise presentation of the disease and its possible effects, outcomes, results and modalities of treatment available. Overplaying, as well as underplaying a given data or the results of investigations, lab results or the observations made, occlusions drawn from the clinical findings, etc. may fall into the category of non-maleficence. Informed consent demands that all the relevant and pertinent information be provided to the patient in simple language, avoiding the technical jargon. It also includes the final outcome, i.e. clear comprehension of the situation by the patient.

As to who should give the consent in a given situation is a matter of simple choice. In a normal, mentally competent adult, it should be the patient himself, and not the next of kin, a relative or an attendant. However, in special situations like the children, the mentally incompetent or some other atypical situations it should be a guardian, or a '*Wali.*' In many an unwanted disagreements, it has often been claimed that the patient or the family did not accept or understand, or believe the information. So one cannot highlight the significance of informed consent enough, particularly in a society riddled with corruption, where everyone claims to be the sole guardian of a patient in a milieu of an emergency. One should be careful in such situations as many a greedy cousins or nephews may wish to bring harm to a relative by taking charge of the scene and exploiting the situation later. In such a case withholding of information may be better than releasing it.

The needs of the society create many ethical dilemmas for the physicians. A common problem is the tension caused by the 'conflict of interest' between a patient and the needs of a society. This dilemma is often the result of a physician's primary responsibility towards his patient and the over-riding principle of promoting the good for a group, a community or a country, based upon the fundamental principle of 'justice' and 'equity' for the group as opposed to an individual.

For instance if an Otologist has a special interest in performing a Cochlear Implant, costing several thousand dollars, he may prefer it over allowing the given amount of money for saving the lives of a large number of children through immunizations. Such a conflict of interest is often described as "Bedside Rationing."

Since a physician has somewhat of a fucidiary approach and bondage with a patient, he becomes biased towards his individual interest ignoring the society at large. Such a form of bedside rationing is evidently unethical. Bias of any kind is bad, but if one has an authority and abuses it towards his bias, that is worse, and potentially harmful to the society.

Oral cancer is the second commonest tumour after the Lung cancer in both males and females in Indo-Pakistan. Most of them present quite late for treatment. In an advanced cancer of the mouth, the choice of treatment is limited. The treatment options for such a patient are radical surgery, radiotherapy, chemotherapy or a combination. The gate keepers and the fund providers, as well as the economists use a terminology, called the "opportunity costs", which includes the benefits from an alternate use of resources in lieu of other services [3].

The parameters for comparing the various modes of therapies are ambiguous. One approach could be, 'measuring the amount of life saved' by a given treatment, and the 'changes in the quality of life' produced by the treatment [4]. While the former is a tangible entity and can be measured on a scale of time, the later is as abstract as entities like fragrance or colour, etc. Nevertheless, experts have developed measuring scales for the quality of life also.

The awkward question is that the numerical assignment of value to various states of health is difficult if not impossible. Administering 'value' surveys to segments of general population has done this. The questions asked are neither easy to neither design nor answer. The examples include what percentage of shorter life would you prefer, with a normal looking health, instead of a longer life with a major disability? [5]. And an equivalence of number of questions such as how many people with chronic non-fatal illness would have to be saved from death to make saving their lives preferable to saving a smaller number of patients in good health?

In a study conducted by McNiel et al. [6] a questionnaire was floated as a tool of research in healthy volunteers, inquiring their preference for total laryngectomy with loss of voice as opposed to radiation therapy with useful voice but perhaps diminished longevity, in advanced laryngeal cancer. Twenty percent of these healthy subjects gave the answer that they would prefer to retain their voice with a shorter span of life with normal voice than have a longer life of a poor quality due to loss of voice. But the ethical question that arises is that how often do we surgeons discuss all the iatrogenic consequences with a patient in our society. Often enough a patient suffering from a deadly cancer of the head and neck region runs from pillar to post to find a hospital bed. For him a soft word from a physician is sufficient reward for his suffering let alone the courage that he may muster to even question the credibility of the surgeon, or the possible outcome of the treatment. McNeils' study suggests that often enough if the ethical formula of informed consent is observed some patients may indeed prefer quality to quantity of life in advanced clinical conditions. Living with honour and dying with dignity is one of the attributes of human rights.

8.1.2 Medical Futility Therapeutic Nihilism

A new ethical question cropping up nowadays is the question of Medical Futility. I recall an episode that a colleague encountered as a Head and Neck surgeon a few years ago. It was a case of advanced facio-maxillary cancer. The patient was an elderly man, whose pathology had completely destroyed his facial structures. Infact he was discarded by his family, and he was somewhat of a destitute. Despite the repeated pleas that there was little one could do, he refused to leave the hospital. Infact at one stage he threatened to commit suicide if no treatment was given to him. The ethical question of autonomy dictates that he had a right to demand treatment, and the physicians were under an oath to provide one. After all life and death belong to Allah and we as physicians are only ordained to do our best. At long last the man was managed with extensive surgery, after which he survived for about a year. The physician concerned knew that the outcome will not be either quality or quantity of life, and he explained this to patient at every examination.

The concept of medical futility is such that in a situation like such, the care provided is going to be futile indeed burdensome, hence the treatment should be withheld, or withdrawn. Some even argue that not only that the physician should abstain from offering futile therapy, but that recommending such form of treatment could be some kind of negligence. The question is that who should make a decision of such a significant nature, and how can one predict or a guarantee the outcome of an illness. After all haven't we seen patients who have practically died following a cardiac arrest, or remained in a coma for years to return to full life and lead a normal life afterwards. The books on what has been described as death-like experience are a sound proof of the first hand encounter of these subjects with death. They survived, and suppose in a situation like this, if the ventilator or other life support systems were withdrawn, presuming the prolongation of treatment to be futile, then what would have been the fate of these people who survived despite all odds.

I recall a case of advanced Verrucous Carcinoma (T4) of the cheek, narrated by a colleague, where for the previous bad experiences of radical surgery, he thought it would be futile to attempt one. But some how, something at the back of his mind kept insisting that some form of treatment must be given. So more as placebo, or perhaps as a palliative measure, he prescribed an anti-viral drug and sent the patient away. A few weeks later the patient returned with an amazing appearance of a normal looking man. If this physician had not known or treated him, he would have disbelieved his eyes.

Hence, the concept of medical futility has to be applied only with great caution. Yes, there might be an instance of incurable Oesophageal cancer, where all treatments may have failed, and a state of terminal illness requiring palliative care may be the only option left. Now if this patient has to be exposed to a feeding gastrostomy, this decision should be based upon human empathy and not for merely for nutritional support thus prolonging his misery. It is a slippery slope, as some may argue that feeding must be continued through a gastrostomy, but others may not agree. If you can't cure it, one may not prolong the misery by delaying the dyeing process. On the other hand one cannot discontinue nutrition and hydration in the expectation that the end would soon come. It may not, and the decision by the physician by withholding a procedure may become the sole cause of his death, in the eyes of the relatives.

The family and the ethics committee must be consulted in all such matters. No one will argue that fluid intake and analgesics and sedatives must be continued till the last, but should active intervention like a gastrostomy for feeding purpose, is something that many experienced physicians may not like to perform.

Futile therapy is considered to be the end of the spectrum of treatment. In centres where this problem is developing fast enough, the decision is jointly made by a group of experts comprising of the physician, the patient, the ethics committee representative and the managers. Palliative care is the end of the road. It has many unhappy questions to answer and certainly cannot be approved in the Muslim societies except for an extremely rare situation, as I have discussed in a separate essay.

From the patient's point of view, medical futility poses a different dilemma. Once the treatment is declared futile, he remains under no moral obligation to continue to suffer. Thus we are witnessing a growing number of terminally ill patients who demand an end to their life. The emerging dilemma of voluntary euthanasia appears to be the tip of the iceberg, in those societies where the love and the care of the kith and kin are absent. So many countries have granted due legitimacy to voluntary euthanasia, that one fears to imagine the development of a sinister state of affairs. After all the suicide doctors have become a reality of life in some developed countries.

In 2008 a devilish doctor who called himself Doctor Death applied for a British visa, to come help dying people expedite their death. Sensibly enough he was refused entry by the home office. But many people travel from the UK to Dignatis in the Netherlands for assisted euthanasia. It is legal in that country but is it ethical? And what of the burden of liability if not sin on the conscience of medical personnel who help in the act.

While folks in the developing countries suffer at the hands of economic constraints, they are fortunate enough to have the close bondage of family units that keep them not only alive indeed cheerful despite all odds. To them the old folks are a treasure, a mine of wisdom, a shady tree under which the younger ones may rest and learn the art of living with the hindsight so easily available to them from their elders. The entire fabric of the Eastern society is like a fine piece of tapestry. One thread holds the other neatly and firmly. Let this irreplaceable fabric remain intact, for this act alone can save many from some unethical decisions. In any event west or east, the moral obligation should he such, that the decisions in medical care of the fatally ill, the elderly and the disabled be based upon 'justice.'

References

1. Dougherty CJ (1991) Ethical problems in health rationing. Health Prog 32–39
2. Yanagihara N (1997) Medical ethics and resource allocations. Panel discussion at IFOS XVIth. World Congress, Sydney, Mar 1997
3. Mensel PT (1992) Some ethical costs of rationing. Law Med Health Care 20:57–66
4. Lantos J et al (1997) The concept of futility. Panel discussion on medical ethics, XVIth. IFOS, World Congress, Sydney, Mar 1997
5. Sessions D (1997) The concept of rationing in medicine. A panel discussion on medical ethics at the XVIth. World Congress, Sydney, Mar 1997
6. McNeil BJ et al (1981) Speech and survival: tradeoffs between quality and quantity of life in laryngeal cancer. N Engl J Med 305:982–987

Quality of Life, Assisted Prolongation of Life, and End of Life Issues

9

9.1 Quality of Life Issues

Savants in the field of health economics have coined some highly meaningful catch-phrases over the past couple of decades. One such term is the impact of disease on human health or, more specifically, the *global burden of disease on health*. It has many nuances and ramifications in a person's working life and indeed on the community in general. No doubt there are a fair number of clinical conditions that cause significant morbidity and incapacity to those who sufferer from it, resulting in loss of work and thus an economic burden on the people's family and society. Some diehards may even cite an example of chronic intractable vasomotor rhinitis, which may indeed affect quality of life, with repeated bouts of sneezing, rhinorrhoea, and nasal stuffiness. But that is only a minor example compared to someone suffering with advanced motor neurone disease or painful sciatica or indeed ageing.

Another interesting terminology used is quality-adjusted life years (QALYs). Determining QALYs uses a complex mathematical formula that health economists employ to arrive at a decision as to whether a given condition and a given treatment would result in x number of QALYs. Vertigo and ageing, are essential contributors the quality of life [1–3].

The QALY has been defined as a measure of burden of disease on the society/community/humanity in both quantitative as well as qualitative terms. It is used by health economists and the 'gatekeepers' as a tool for measuring the cost of a medical intervention. It is calculated on the basis of number of years of life that would be added by a given intervention. The value given to a year in life is 1 and to death 0. For example, if patient loses a part of his body – an organ, an eye or limb – and cannot maintain a normal healthy life without wheelchair support, then the extra life years are calculated between 0 and 1 to account for it. It is a measure of cost utility designed to calculate the ratio of cost to QALYs saved for a particular intervention. It is then used as parameter to decide the distribution of health resources. Rationing in today's health care system has become the norm; hence all the debate that goes on in the media as to when and why one patient should be preferred over another if everyone is contributing equally to the service.

S.H. Zaidi, *Ethics in Medicine*,
DOI 10.1007/978-3-319-01044-1_9, © Springer International Publishing Switzerland 2014

The tools employed for QALYS are time tradeoff, standard gamble and visual analogue scale. Further details can be obtained through health economics and medical ethical journals.

Quality of life is currently an important parameter that seems to interest health economists, and truly enough it has a major role to play in one's life. Perhaps QALYs would be more applicable in cases of intractable or incurable illnesses. Nevertheless, it is a measure that determines the application and outcome of a given health service and distribution of resources depending on their availability.

The reader is referred to a wonderful article written by Hassan et al. [4] on the subject of the psychobiology of posttraumatic stress disorder . It is a common practice to see patents with chronic illnesses suffering with loneliness, depression, anxiety, emotional stress and lack of physical activity. It is the perpetual fear of pain and suffering, handicap, or loss of independence that seems to trigger the initial fear, which then takes the shape and form of a vicious cycle. Such patients have a tendency to give up work, which means going on benefits (in a social welfare state like the UK or the USA) or go begging. Another source of burden on society is the loneliness due to the loss of social activity, lack of communication with other people, and a tendency to curl up in a cocoon of depression, solitude and misery.

In most countries, the burden of disease and disability is borne by the taxpayer; however, in many countries that lack social services, this burden is happily shared by the family. In the Asian communities of India, Pakistan, Bangladesh, the Middle East, and many Far Eastern countries it is the norm to have large families sharing home and hearth. In Chinese communities, too, this phenomenon is witnessed – not just in their homeland but all over the world wherever they may live. It is all a matter of cultural norms. Altruism is a universal practice, which is perhaps more visible in Asian culture than in the West.

The effect of culture on mental and psychological health is an established and highly debated subject. There are certain culture-specific psychiatric syndromes such as *Koro* and *latah* (*Silva de Pdamal*) [5]. The term *culture* has been defined by sociologists and anthropologists in many ways. Culture may mean different things to different people. It includes, for example, values, norms, practices, faith, beliefs, habits, family patterns, upbringing, taboos, sanctions and dietary habits.

It is a common observation that in many developing nations the family bond is so strong that the burden of disease and disability does not affect the society as it does in the developed nations. Because of economic reasons, though, it is regrettable that these values are somewhat fading away in the Asian diasporas now living in the West.

Modern life is full of trauma, fear, anxiety, distress, warfare and misery. Posttraumatic stress disorder is a relatively new diagnostic category, but it carries enormous significance in describing or managing the effect of a traumatic condition on a human being [6]. Although some famous politicians in recent times denied the existence of 'society' in their political speeches, most sociologists would disagree and stress that the society or a community is indeed a collection of many individuals. Therefore it is illogical to assume that the impact of a traumatic situation affecting an individual may indeed be reflected in the collective response of the community.

In the bygone days when mankind lived in the forests and had to hunt for his meal, tribal culture developed in response to the collective needs of time such as protection and safety, etc. Even today in many parts of the developing countries a tribal head, who is usually less than a kind person, is more revered than the king emperor and the tribal customs are more valuable than any legal or religious commandment. History informs us that even though the prophets, savants and the saviors came to serve the divine commands, more often than not they had to accept and imbibe the cultural norms of the society and only gradually if at all, discard them over a period of time.

Unfortunately the eastern families living in Britain and other parts of the west have, for economic reasons become equally dependant on the social and welfare services as the families have become nuclear.

Old age is often associated with loneliness, solitude, isolation and dependence on others. Ask any old person and he will say that independence is a gift of God, and dependence a curse. No old-age pensioner wants to live on the meagre help of kin and family if he can avoid it. Hunan dignity and honor are duly compromised, even if one has done all that he could for his wards but is now obliged to be a burden on them.

One wise man said that he prays to Allah to beg Him to take me way while I am still able to walk on my two feet and take care of my personal needs. Nothing is worse than old age in terms of dependence on others. There are examples of children who have abandoned their elderly either for want of resources or mere neglect!

Imam Sajjad pleads to Allah in one of his supplications to grant adequate means in old age so that one does not become dependent on anyone other than the Almighty.

9.2 Assisted Life Prolongation

This is a common dilemma faced by the family of a patient needing artificial means to prolong life and the medical profession. Modern technology has made it possible to maintain life per se as long as resources allow or there is a need to maintain such a life. It may not be a life as we know it, but life nonetheless, although it is usually called a vegetative state.

In a pluralistic society like Britain, it is a common dilemma when there is an unfortunate accident, cardiac arrest or some other reason for a patient to require cardiorespiratory assistance. The decision is based upon the needs of the patient and not the availability of resources. Islam clearly mandates that *Ajal* is inevitable, for which there is a fixed time, and cannot be postponed. So if a Muslim patient requires such artificial prolongation, should it be carried our or not?

The fundamental question that arises is the reason for doing so. Is it for a family reason, as is the common practice in most Asian Muslim cultures to wait for the relatives to arrive to share the burial. Islam clearly mandates that the burial must not be delayed.

Therefore, as a Muslim, it is essential to remember that there be no undue delay in the disposal of a body. So, if a patient has sustained an incurable clinical condition, a trauma, cancer, terminal illness, or coma, where the physicians have no hope of recovery with a normal life, then the question of artificial prolongation of life becomes a practical and challenging task. Each case must be assessed on its merit by the team of physicians and with full information provided to and discussion with the next of kin.

No doubt, autonomy dictates that all human beings must be treated equally and have equal rights for life and full treatment. It twice demands that every thing possible must be given to them, but whether these services would be of value to the patient in a situation as mentioned above is the question that has to be answered by the physicians and the family together. It is the decision to attach such a patient to a ventilator that is more important: once attached, the decision to disconnect may pose an even more intense dilemma.

In countries where there is no social system and a family has to bear the cost of prolonged life assistance, the dilemma becomes manifold. Emotions at the time would dictate that the patient be supported artificially, but as the days change into weeks and months the family's resources run out, leading to many untold miseries for the rest of the family members. Here the physician plays a pivotal role: he should give his most honest detriment and skilful opinion about the future outcome of the illness. Perhaps one physician alone may not be the best answer, so either a team of physicians involved in the patient's care or an ethics committee may decide, keeping all the resources, both financial and medical, in view.

One learned *Alim* was asked this question in the USA, where health care is based upon one's contribution to insurance companies. His answer was that the patient may be kept on a ventilator as long as the insurance company allows the hospital to continue the service. So when the resources run out, who should make the decision to disconnect the ventilator?.

What if someone has no access to health insurance? What if a family has no resources, and what if the state does not take the responsibility or indeed has no a facilities available?

So each case has to be decided on the facts and not fiction, reason and not emotion. No doubt life is on loan to us from Allah, and we all have to return to Him, but we must respect life as indeed the dignity of death.

Once the patient has been connected to a ventilator, the decision to disconnect is rather difficult; as I have seen in my professional life, such patients whose ventilator was about to be switched off and they showed a twitch of the facial muscle or a movement of the toe on the very morning around when the physician's hand was inching toward the ventilator switch that would ensure his death. That particular patient not just recovered fully but went to the USA to become a top professor.

Recently a colleague told me the story of her brother who was crushed to near death in a fatal motor accident. Emergency services rescued him and put him on a ventilator. After a few weeks, when the physicians had established that he had no chance of recovery and asked a colleague to inform the family that they were going to switch off the ventilator that afternoon, she said that she pleaded for one more day. She told me that it cannot be anything except the miracle of Maula Abbas AS. The next morning they were all ready to see their dear one die, and as the nurse appeared at the switch to turn it off, they saw the young man move his toe just a wee bit. They all stopped and waited. He recovered fully and is now a renowned aeronautical engineer.

These and many other examples inform us that no one can determine the time of death except the Almighty. We cannot either expedite nor delay it, even for as short a moment as the blinking of an eye.

Switching off a ventilator should depend on the status of the patient, that is, confirmation of brain death, as discussed elsewhere, although some societies and faiths do not agree with that criteria. An ethics committee is perhaps the best authority to decide in such situations. Collective responsibility at least eliminates an individual bias, and collective wisdom is usually better than an individual's decision

An interesting principle mentioned by some ethicists is called the 'principle of proportionality.' It basically means that the fundamental role of a physician is to provide relief for and benefit to the patient and such treatment is therefore sufficient to outweigh the burdens of responsibility. It is of particular relevance in such situations as described above, with such a resuscitation with assistance through intervention, in which many factors should be considered, namely the potential outcome and benefits of resuscitation, which may actually mean not just saving a life but also giving a sense of closure to and the resolution of a sense of emotional or psychological guilt on the part of the family and other relatives, and such major factors as the financial impact on the family, friends or relatives, especially if the prognosis is guarded and the life assistance may involve not weeks but months or even years of care, or indeed resuscitation to a less than desirable quality of life. Medical futility may also be kept in view, which basically means that an intervention or a physician's action may fail to achieve the desired goal, such as active, productive, useful, normal life, totally independent of any assistance.

In matters of resuscitation, it is essential to recall all those principles that have been discussed in some of the essays, namely the issue of medical futility, withdrawing or withholding treatment, an advanced directive, a written living will, family commitment and involvement in decision making, any DNR order and above all communication skills.

In the current environment of limited resources, if the ventilator is required for a young a patient while another older patient waits to be connected, the decision becomes harder. No one can say that the young man may live longer and be more productive in life than the older person, or vice versa. Each individual has equal rights to live honourably. Distributive justice would dictate that both these patients must be given equal opportunities to live, assisted or otherwise, but restricted resources may determine that non-availability of resources may leave the decision in the hands of the physician, who must decide on merit and the voice of his conscience.

9.3   End of Life Issues

TLC and DNR are the two terms often seen written on a clipboard hanging at the bedside of a patient in a palliative care ward in the UK.

TLC does not stand for Total Leukocyte Count as one might guess but for Tender, Loving Care, implying that the patient has been declared incurable and terminal. Besides, DNR is a standing order, which means Do Not Resuscitate is often seen signed by the patient or the next of kin. This essay critically evaluates some 'end of life' issues.

Medical ethics is as inseparable from medical profession. And yet, how little one is taught at a medical school and how sparse one's knowledge is in the practical life as a practicing physician. The maze of ethics is complex, puzzling, beguiling and prismatic, indeed.

Culture determines many aspects of end of life issues. Traditions and cultural norms are akin to a body within which resides the soul in the form of faith, religion and beliefs. These rituals and beliefs help the patients and their close ones in many ways, in the hour of anxiety, grief and bereavement. They provide a context of meaning and a fabric of support; more like scaffolding upon which the future blocks of life are delicately balanced.

As physicians we are obliged to observe the code of conduct taken as a professional oath, and pay due respect to the cultural, religious and traditional ways that a certain community, or an individual observes. Customary diversity seen in our patients adds to the challenges that a physician must meet in his life. Deliberations upon such diversities, as we often see in today's multiethnic and multicultural societies can make manifest latent cultural differences, overcome subtle mistrust brought about by either apathy or ignorance, sometimes racism, and less often the worst of human virtues, 'fear.'

A physician's quality of empathy may indeed become questionable in stressful situations; if he is not cognizant of the cultural variations. Such matters as artificial prolongation of life, a living will or the practice of volunteerism or altruism in matters of organ harvesting are culture sensitive. The Eastern culture is strong textured compared to the Western culture, which is weak and individualistic. So the customs and values strongly differ.

Matters related with terminal illness, care of the elderly, the disabled and those approaching death are all culture and faith sensitive.

Informed consent is the very spine of the principle of autonomy. In the United Kingdom, the individual concerned is the only person who makes the decision; and rightly so as he is the one who must be fully informed about his disease, its diagnosis and management etc.

In most Asian cultures it is often the reverse. If a patient is brought in for say a suspected cancer, more often than not the closest relative would first enter and request the physician not disclose the nature of illness to the patient basically to save him from a shock of hearing bad news. And that is a good idea. But is it ethical?

How often do we have to conceal the truth from the patient only to allay the fears or to retain the trust of the family, at the cost of breach of the principle of confidentiality? Informed consent in many such cultures needs to be redefined, as it invariably involves more than the patient alone. The whole fabric of these societies is woven very closely and quite intricately. Tribal culture has a lace-like structure; delicate threads woven into each other.

In Malaysia, I had quite different experiences. If the patient is Chinese, often enough of Buddhist faith, sometimes, Shinto or less often Christian, it is always the male member of the family who must be fully informed irrespective of the patient's age or gender. The male member would determine the fate of the patient.

If it is an Indian patient, either a Hindu or a Sikh, the full story has to be narrated to the entire family, before any major decision can be taken. Mostly it is the oldest member of the family who must be taken into confidence before any intervention can be carried out. He determines the fate of the patient.

If the patient is ethnic Malay, the story is fascinating. Even if the male patient has to undergo as radical a procedure as Total Laryngectomy, involving the permanent loss of voice, the decision is made by the lady of the house. In fact it must be the only culture where the male members of the society shift with parents of the wedded wife to live happily ever after!

Another example is that of the care of the elderly and the terminally ill. In the UK most elderly are left to the mercy of the caretakers in the old peoples homes. In Asian cultures the elderly are not just loved, indeed revered, by all family members. So when it comes to the question of the care of the terminally ill, matters are culture sensitive and custom based. Perhaps in England a physician may employ the principle of 'Double Effect' of the opiates in the care of a 85-year old man suffering from terminal Thyroid cancer. In rural Pakistan one could not even imagine doing such a thing as not to actively resuscitate this patient through a Tracheotomy for relief of respiratory obstruction, even if futile!. The concepts of Tender, Loving Care (TLC) or Do Not Resuscitate (DNR) are alien to the eastern culture. And yet, a physician has to confront some hopeless situations where he knows that death could be a blessing!

There is no consensus amongst the Muslim physicians on the issue of DNR orders. In one debate which was televised on a British channel in 2012, I was invited to a discussion with a learned *Alim* and a consultant urologist. My opinion was that DNR orders may be necessary to comply in certain hopeless situations like a cancer or terminal illness as prolonging the misery if a physician can't cure is not very compassionate act, but my urologist colleague vehemently disagreed and said that in his illustrious career spread over several years in Ireland he had never signed a DNR form. He quoted several illustrations where patients had survived after a disaster and since he did not carry out a DNR order and helped resuscitate them, and they survived to live on. My argument was that prolongation of a miserable life may not be an ideal scenario, but one cannot and must not assist a process of dying either. The Maulana said that '*Ajal* is inevitable but doctors must continue to strive till the last breath'.

A colleague in the USA holding a top position in neurosciences in an Ivy League university, told me his personal story recently. His father suffered a stroke, choked and collapsed. The ambulance arrived in time and commenced the CPR forthwith. The family kept pleading to the team to let him go peacefully as he had suffered for many years with Alzeihmeres and dementia. But they could not stop the resuscitation efforts as there was no living will or an advanced directive to act on the principle of DNR. This colleague said that "we cried our heart out, begging and pleading, but they were just doing their duty, in other words saving their skin," while the old gentleman suffered and eventually passed away in utter agony.

A few years ago, I was called to attend to a patient with terminal cancer of the upper respiratory tract. He was gasping for breath, and I was asked to perform an emergency tracheostomy, to relieve the respiratory obstruction as well as to connect him to a ventilator. We hastily prepared for it all, when his next of kin showed us the DNR form that old man had signed only a few days ago. Obviously the procedure was postponed in respect to the DNR orders.

Many years ago I was involved in the care of a very famous politician who was in coma for several weeks, and the anaesthetist wanted me to do a tracheostomy, for assisted life prolongation. In the developing countries, dealing with a celebrity is not easy as the repercussions could be positive as well as adverse and there is no concept of DNR or DNAR, over there. A doctor is solely responsible for everything. So I decided to discuss the issue with the patient's family, who were very reasonable. The eldest son decided in favour of abstaining from doing any gross thing such as making a hole in the patient's neck, but the younger son was adamant to proceed. It was a frightful dilemma for me as knowing the culture as I did, it was a very tricky situation. The elder son was possibly keen for the old man to pass away so that he could inherit the *gaddi* and the authority as well as the wealth. The rest of the family, probably genuinely wanted to do whatever possible to prolong his life. Outside the hospital ferocious looking gun men stood guarding us all. Well, we decided to go ahead with the tracheotomy, and prepared to face the worst. If only we had a DNR form, the situation would be different. So just as we were ready to proceed the younger son and others decided to let the old man go who had been in a coma and oblivious of the world for weeks already and to make him more miserable was against their faith.

Writing a will is a routine practice in most countries. A 'Living will' or an advanced directive, on the other hand is not a routine practice, at least in the Eastern culture. But in matters of disposal of the body and other religious rituals, at least verbal instructions are given to the family in the Muslims living anywhere; albeit seldom in matters of permission to harvest the organs. A verbal rather than written instruction is of course acceptable to the religious scholars, but not so in the eyes of law. A living will or an advanced directive at least for those who live in the west should be encouraged so as to avoid any last time delays in the disposal of the body due to official form filling and other matters by the local authorities. It also applies to permission to harvest the organs. Whether or not an individual wishes to write a will for assisted life prolongation is a personal decision. Informed decision may be made by the individual, or the next of kins in such matters. Some scholars believe that suffering is a way of salvation, but that may be left to the Ulema to investigate further. Allah who is Rahman and Rahim may not wish to see His humble servants suffer, when they can be helped! The basic decision to provide assisted life support through machines must be made with great care and much discussion with the family, and the next of kin. Factors such as financial liability and resources, etc. play a significant part in such a decision making. In the UK and USA it may not be a major issue, though it has great implications here too, but in India, Pakistan, Banagla Desh and many other Asian countries where the family has to pay for all such expenses, the decision becomes hugely difficult. No one except the immediate family may be able to assess their situation. Obviously if they can't afford it, they should take a decision accordingly, and not put the patient on a life support system, it is particularly so in fatal and terminal conditions. Once attached to a ventilator or life support systems, to switch it off or disconnect is a huge responsibility, and is akin to euthanasia in the eyes of many moralists. Of course once brain and somatic death has been confirmed, the life support system would be withdrawn.

Here I would like to quote an example of a very famous Indian surgeon, who left a living will which was a model for others to follow. He was a Sindhi Hindu, who visited his home town, Thatta at the mature age of 80, which he had longed to visit since he left as a young boy at the time of Indian partition. He was simply overwhelmed by the experience and later on published his visit in the illustrious *Hindustan Times*. I was fortunate enough to be his host, so apart from those publications he also sent me a copy of his will. He had clearly mandated in the will, against undue efforts to revive him in his fatal illness and to remove his organs in time to donate to Muslims and Hindus or Sikhs or Christian, whoever could benefit without any racial religious or sectarian prejudice, purely based upon the urgency of the need of the recipient.

Many academic institutions and donor agencies involve the scientists of the developing countries, in conducting clinical trials, which they may not be allowed in their own countries. Many times, when a missionary goes to a village in Asia or in Africa, where even the drinking water is not available let alone medicines, he is taken to be a God's messenger to them. Hence whether he treats them as his patients or as his guineapigs is entirely unto his conscience. Often enough the natives can't even understand his language, let alone think of obtaining informed consent; and the less said of the interpreter the better. More often than not he is a paid employee!

On the subject of 'human guineapigs' a renowned physician, blew the whistle when I was a humble student at London university in late 60s. I do not recall it all, but I remember that the medical profession had disowned him hastily along with his published work. He became somewhat of a recluse. Those were early days and medical ethics was not even recognised as a teachable entity! In some developing countries, the prisoners and the inmates are often obliged to donate blood for transfusion, without their consent. What happened in China recently is another example of human apathy. The prisoners condemned to death were quickly whisked away after the execution and their organs harvested to be sold out in the black market. In an Asian country it is often reported that many poor people sell their kidneys to the well to do, obviously through the connivance of the medical professionals. In Pakistan, the national press often advertises for any person who wants to sell off his kidney, for a given sum of money. It has often been suspected that the surgeons involved are not quite concerned about the tissue matching and such time wasting intricacies of science. After all one is all but saving a life!

Some time ago, the British TV channels showed a father complaining of the arrival of his dead son's body from the holiday resort of Costa Rica, where he had gone surfing, with all his organs removed.

So if that is what is happening in the most civilised societies of the world, I dread to imagine what all that must be happening in our countries where the knowledge of medical ethics is almost non-existent and human rights openly violated. But does this entire philosophical sermon really matter in a society where innocent physicians are perpetually being killed, molested or forced to migrate!

Therapeutic nihilism and futility of treatment is a burning ethical issue these days. It basically means that in a given situation, medical treatment may be considered to be futile and the competent authority may decide to abandon the treatment and let the patient succumb.

Medical futility is an abstract but many times a tangible entity. It is an important consideration in terminally ill patients as indeed sudden tragedies where the organs may be poor, and the physician feels that an intervention will not bring about desirable outcomes.

Decision to withhold treatment or withdraw life sustaining equipment will depend on the basic decision on medical futility. So it is an important ethical principle that should be given due respect and authenticity.

Much debate revolves around the definition of medical futility, and factors that may indeed be the judgment call from health delivery staff rather than true and genuine futility of treatment. Besides there is always the fear and concern basically if the family that medical futility may deprive the patient of a potential benefit that may arise from somewhere, even as a miracle.

But the most important fear of the application of medical futility bill is the factor of saving the cost, on the part of health authorities, especially if it is provided by the social welfare state or an insurance company. As some fund manager may sincerely push the medical profession feel to save the money in such situations for other patients who may have better chances of survival and achieving the quality of life.

The million dollar question remains who should make the final decision of ending a life or prolonging it through artificial means?

General advice is that in all honesty and with a clear conscience if the physician or the team of physicians have taken all factors into consideration and if is considered in these situations that further treatment will be futile and therefore the dying process may not be delayed, they should inform the next of kin with utmost empathy, confidence and sincerity and help them reach a definite decision.

Most religions do not consider death as annihilation or failure. Every living soul in the universe has to die. It is the death with dignity that is more important in the contemporary world than dying perse.

Indian subcontinent is a melting pot of many faiths and philosophies. Hindus believe that '*atma*' is the soul, which is a gift of *Eshwar*, and is the final truth and ultimate reality. *Brahmans* genuinely believe that a good death (*su-mrtyu*) will come at the most appropriate astrological time and at the right place. Muslims have clear concept that every living being has to taste death, and one must be prepared for it at any time and any place.

Buddhists live in many countries in South Asia, they believe that death is ubiquitous and inevitable, and accept death with peace and dignity. They find peace in mediation which develops their immunity and increases their inner strength to face adverse conditions with courage, patience and inevitability. Some may refuse to take even analgesics and strong sedatives. To reject and abandon patients in a less than respectable, i.e. vegetative life and withdraw even the nourishment is according to Budddhism, a denial of the basic human values of compassion and mercy.

Judaism is an ancient religion, which has many principles common to Islam such as circumcision, or praying to one single God. Judaism believes that life has to end in death, and the human body belongs to God almighty, but we must avail of treatment available to us to eradicate illness and take good care of the body as it is on loan to us. Suicide or euthanasia is not allowed in Judaism. On life assistance the Jewish concept restricts the permission to withhold or withdraw treatment in conditions where the physicians assume the inevitable death in 72 hours. As soon as the person is diagnosed with fatal illness (a *terefah*) it is allowed to withhold or withdraw treatment, or disconnect the machines, if it is in the best interest of the patient. Many conservative rabbis even discourage the continuation of nutrition or hydration if the death is assumed to be inevitable such as in palliative care for fatal illnesses. They believe that dyeing process may not be prolonged and nutrition may be just doing that [7].

Christian thought is deeply influenced by culture. It is therefore dependant on many factors, however, most believe that there is no obligation to delay death, but there may be a duty to employ to use technology to gain one more chance to ask forgiveness of those who may have been harmed. To save life at all costs, however, is condemned by most Christians. Analgesics and sedatives without compromising consciousness is allowed in terminal illness and in palliative care. Catholics, just like Muslims also believe the life to be a gift of God and we are only stewards not owners of our bodies. In the light of the teachings of Jesus Christ and his suffering, good Christians acknowledge pain and suffering as indeed death as a way of penance and relief from agonies of this world.

Protestants are happy with the life sustaining machines, and agree to the withdrawal of machines in certain inevitable situations. They oppose Euthanasia and assisted suicide, but they also do not encourage doctors to play God. There may be some though, who may show affirmative approach towards physician assisted suicide in very specific circumstances. Some Catholics also show similar inclinations.

In the Protestant Christianity, no general consensus has been reached about continuation of hydration, nutrition, etc. in a patient in a vegetative state. So it is all up to the next of kin to determine the course of action.

Islam is the final divine faith. Death is inevitable, but should it be delayed artificially; it is a debatable issue.

Zahedi [7], quotes various Muslim philosophers like Farabi Avicenna, Mulla Sadra who have discussed *'essentialism'* and *'existential transformation'* of man towards *'perfection'* and *'transcendence;'* and anyone who has not regarded death as non-existence and imperfection should rethink. Every thing in nature is recycled. No one is perfect except Allah.

Rumi has described the cycle of life in a beautiful stanza included in an essay in this book and only a few years ago in a book edited by Dawkins, a renowned scientist has described the life cycle of carbon, also included in this book, both reaching the same conclusion, i.e. death is a bridge between life here and hereafter.

Keeping the philosophical debate of such greats like Mulla Sadra in view and more importantly the teachings of Quran and Ahlebeit every Muslim knows that the life is precious and must be taken care of. But death is an ultimate truth. And one cannot delay it at the final call. God exists and is immortal, the rest of everything anywhere in the universe, space, oceans, space or beyond must perish.

Saving life is an *ibadat, and a* physician must endeavour to prevent premature death. Muslim *fuqaha*. Unanimously agree *that* decision to intervene to prolong life must be thought carefully, but once an invasive intervention is commenced, the life saving equipment cannot be turned off, unless the physicians are absolutely certain about inevitability of death. They also agree that treatment does not have to be provided if it merely prolongs the final stages of a terminal illness as opposed to treating a superimposed, life threatening condition.

Zahedi [7] discussed the issue of pain and agony in a philosophical way. He quotes Quran, with reference to pain as a form of trial "O all you who believe, seek assistance through patience and prayer, surely Allah is with the patient … Surely we will try you with something of fear, and hunger and diminution of goods, lives and fruits; yet give good tidings to the patient who, when a misfortune befalls them, say 'surely we belong to Allah, and to Him we return;' upon those test blessings and mercy from their Lord and those; they are truly guided (2: 153–157)".

Pain, according to the perception of some Muslim scholars is a means of self purification with an educational and spiritual purpose, a means of self purification from a sinful life. Pain, however, according to many scholars, be relieved with analgesics until the time of death.

In the current environment of shrinking resources and rapidly increasing costs, the fund managers plan out certain places that may determine the final outcome of the therapy or treatment given to such patients.

Islam does not permit any form of suicide. Euthanasia is one such exercise, which is not allowed in most societies. In the West, currently a debate is underway, whether or not assisted suicide may be allowed in some desperate situations. Netherlands has a special clinic, where 'death with dignity' is the slogan. In the state of Oregan assisted suicide is allowed, and some other states are considering it too.

UK is very conservative and sensitive on this issue. Although the population of church goers has shrunk over the years and religion is no more a dominating force, public still maintain some form of quiet dignity in the matters of life and death. Not a single day goes by that some form of protest is witnessed on the streets of London, Birmingham etc. against abortion, or similar controversial issues. But each day voluntary euthanasia is also becoming close to recognition. Why? One might ask.

Well, there is not a single answer to this question. In the western countries, the society is breaking apart. The old fashioned family units are almost dead, and the faith believers have all but become extinct. What is worse, is that the compassion and charity that is so often seen on the media in natural disasters, etc. does not exist when it comes to the family units. Teenage pregnancy, and single parent families are common in the UK, and Europe. The measuring rod for moral standards like respect for the elderly or care of the young ones is becoming a rarity. As soon as a daughter or son is able to stand up on his own feet, he is pushed out of the parents' home. Then he or she is on her own. Whatever the misery or hardships he or she faces is no more a parents' worry. There is no family support or indeed any help either financial or psychological, or physical available to them.

So if a person grows up in isolation, then obviously he becomes oblivious of others needs, in a very selfish mode.

An elderly or disabled person in the west is a liability to both the family and the state. Therefore one is not surprised to hear of all those heartrending stories of such people thrown out to the mercy of the caretakers and the nursing homes. Some of them end up so miserably that they desire to die.

A totally divergent picture is seen in the eastern cultures, particularly in the Muslim countries. Family values and traditions dominate their culture. Elderly are treated with respect and dignity. They are respected for their wisdom and approached by the younger family members in all matters for advice and guidance. Best of all they are loved for all the sacrifices they made to see the family prosper. The entire Muslim culture is full of stories of altruism and sacrifice. So the concept of leaving the sick and the elderly unattended or at mercy of the caretakers is extremely unusual. If at all it happens as I have seen in some cases in England then it is either the family has no monetary resources to employ even part-time help, and everyone has full-time employment or there may be an uncompromising and obstinate daughter-in-law or indeed son-in-law, who is inhuman to say the least, that the elderly are put in a care home. Normally, it would not happen in Pakistan, Malaysia or any of the Muslim countries.

Sadly though, many horrible stories are beginning to emerge even in Pakistan. Last Ramadan, a TV anchorperson interviewed several elderly women, whose sons had left them to the care of a charity home run by the angelic Maulana Edhi. The sons had simply dumped their mothers and disappeared either due to economic reasons or at the behest of a nasty wife. How awful!

Regrettably the debate on futility of treatment and therapeutic nihilism is more academic than practical in those countries. In the West for the Muslim population once again it is unusual except that the treating physician may decide that the patient has reached the end of treatment line and there is nothing else that can be done, and the patient may then need to be transferred to a hospice. In such situations despite all efforts and utmost honesty the family may have no choice but to surrender.

In the Western society many a question are being raised on the expenses carried out on a patient who may not recover from the illness; thus it would be a case of terminal illness.

Rising cost of treatment in the NHS as indeed elsewhere has forced the health authorities to ration out the services. Therapeutic nihilism is one of the items on the agenda. It means that once it has been determined that further treatment would be futile, a plan of action would be set in, which will let the patient die peacefully. Liverpool care pathway is one such example,; now under review.

People with different faiths live in today's pluralistic Britain. Due respect to their faiths is given in the care of the patients. The Jewish believe that the dying process may not be delayed but the patient must be supported with fluids and analgesics as well as opiates. Most other faiths follow more or less as commanded by the treating physician. Muslim patients either move such a patient to their homes for the final end, or continue to support the terminal patient with prayers and supplications until the last minute. *Quran* has mandated that '*Kulu Nafsin Zaiqatul maut,*' so we recite *InaLillah Wa InLaillah wa Ilehe Rajaoon*. And surrender to the will of God.

It's a documented and an established practice in all Muslim countries that any form of euthanasia must not be carried out. Sanctity of life is duly highlighted in *Quran*: "Do not take life, which Allah has made sacred, except in the course of a just cause (17:33)" and the famous and oft quoted Ayat "If anyone kills a person, it would be as if he has killed the entire mankind (5:32)." Regrettably the exact words of *Quran* cannot be translated in any language to its fullest meaning, and the translation is always less powerful than the actual Arabic words in *Quran*.

Likewise on the matters related to mortality of life and inevitability of death *Quran* is very precise. It says "When the time (of death) arrives, they cannot delay it for an hour nor can they bring it closer by a single hour (16:61)."

At another place *Quran* dictates and reminds us of our helplessness in the matters of life and death, when it says "And no one can ever die except by Allah's permission and at an appointed time (3:145)."

And then in Surae Luqman (32:34) we are told that we have no control on the time or the place where and when we will be called back.

So, one has to recall the philosophy of *Qaza o Qadar*, (destiny and fate) and accept that there are certain things in life that we are obliged and have the authority to perform with free will, but there are more salient acts on which we have no control. They will happen though you may get some respite, mainly on account of your good deeds. But inevitability of certain acts in life is the fundamental principle underlying the concept of *Qaza o Qadar. They* go parri-passu with the notion of *Jabro O Ikhtiar.*

As far as the end of life issues are concerned, the Muslim *Ummah* is struggling to reach a unified consensus on some major issues, but *Fatwas* are now available in both Sunni and Shia traditions about many other elements. For instance Sheikh Abdul Aziz bin Abdullah bin Baz has clearly mandated that euthanasia is forbidden in Islam, including the removal of the life support system in a comatose patient. He said that it was against the *sharia* to decide regarding the death of a person before he is actually dead. He said that life of any person cannot be taken for any reason (http//www.islamicvoice,com,july 97/NEWS.HTM).

The famous Egyptian Mufti Shaikh Yousef -Al-Qradwi has called euthanasia no less than a murder but has allowed withholding of treatment which is deemed useless [8].

Marjae Taqleed Ayatullah Syed Khamenei has called euthanasia *'haram'* in all its forms; and Ayatullah Nuri Hamdanai also regards all forms of euthanasia totally 'haram' [9].

Islam forbids euthanasia, but there may be a situation where the doctors have done their utmost and failed to cure a patient. He is now in state of vegetative life, needing long-term life support. Should the process of active intervention continue or not? And that is a present day dilemma that the learned scholars have to solve. There is a wide and varied degree of opinion variation on this issue. Let us be clear that we are not talking about a situation where there is even the tiniest hope of recovery but those hopeless situations where mortality of the physician glares into their eyes and they simply cannot do any more than wait for the nature to take its course. But since the patient is on a life support system, the death indeed dyeing process may be unduly prolonged. So what is the mandate on that?

At the Najaf meeting in April 2012, this question was raised by a neurosurgeon to a panel of learned Ulema and many physicians. The Ulema continued to hammer the point that "the doctor must continue to save the life." It was very difficult to convince them that the physicians had left no stone unturned to save the life. The time had now reached when doctors were helpless and could do no more. Eventually the physician on the panel had to explain to the audience the medical definitions of death as applied in day-to-day practice. Once the ethics committee and the panel of experts have determined that the patient is brain dead, the concept of futility of treatment steps in. The services now being provided to this brain-dead patient could be employed to serve another patient who may have a normal life with a good quality, instead of delaying the dyeing process of the brain dead. Besides, after the patient has been declared brain dead, time is of essence in order to harvest the organs for saving other lives. The definition of death of Qalb, Fuad and Lua'ab are discussed in Medicalethics.imamaimedics.com; 2013, Vol.2:1.

The Islamic code of Medical Ethics in their first international conference held in Kuwait in 1981 had declared that "If it is scientifically certain that the life cannot be restored, then it is futile to diligently keep the patient in a vegetative state by heroic means or to preserve the patient by deep freezing, or other artificial methods. It is the process of life that the doctor aims to maintain and not the process of dying. In any case the doctor shall not take a positive measure to terminate the patient's life [10]."

Islam has clearly taught its followers that *'Ajal'* is inevitable and all mortals have to return to the Creator upon call. One can neither hasten nor delay the time determined by *Allah Subahanhu*. Death is the ultimate truth that every living creature has to taste. As said in Quran: "*Kulu Nafsatun Zaiqatul maut.*" Therefore as Imam Ali As said, "remember your death every day, so that you do not become oblivious of it." In fact it is widely known that Imam Ali spent considerable time in the graveyards. We are reminded also that when you cross a graveyard do not forget to say "*assalamo alik ya ahlal qabbor.*" So death is inevitable and we must keep it in the forefront.

When it comes to the question of defining death, many terms come into focus. Medically speaking the commonest terms are:

1. Somatic or Biological death, and
2. Brain death.

By somatic death it is meant that the cardiopulmonary and hepato-renal systems as indeed the biological growth of the bodily tissues cease to function, thus resulting in death. The brain death is more complex to define. It basically means that all forms of neurogenic activity in the brain and the nervous system must cease to be labelled as brain death. In other words, it means the irreversible cessation of higher brain function, i.e. cerebral hemispheres, with preserved brain stem function, that would allow the procurement of organs within a specific timeframe.

It is a common observation that in the ICUs, where a patient may remain attached to a ventilator in a state of coma, may continue to grow his hair, nails, and his inner organs. In fact, comatose patients have gone on to deliver a healthy baby.

So although theoretically he may be dead he is not really dead. In the BLS course the instructors always use a phrase that "one is not dead until dead cold." A warm dead is not quite dead yet, as the circulation is still continuing. The life support system ensures just that.

Therefore many ethical questions crop up, such as how to determine the proper time of death so that organs may be harvested. Brain death is the final death as a matter of fact, and in many countries a formal protocol has been laid down by the authorities to establish the real brain death. Brain death is best defined as "complete and total stoppage of brain and brain stem functions synchronously." Such a patient would show total apnea, completely absent brain stem reflexes and total failure to elicit any form of cerebral response or reflexes.

The decision should be made by a team of experts including a neurologist, an anaesthetist, an ICU person, the primary physician, a forensic physician, the specialist nurse and representative members of the local ethics committee including a religious authority. They would then run through the protocol, such as giving 100 percent oxygen and then disconnecting the ventilator to monitor any sign of life through clinical evaluation (as is taught in the BLS courses to all physicians, i.e., look, listen and feel). The clinical tests like the pupillary reaction to light, the Plantar and other bodily reflexes, and EEG and ECG are then recorded. The patient is the given 6 L/min of Oxygen and allowed to build up CO_2 to 60 mmHg so as to stimulate the brain through hypoxia and hypercarbia, while recording the EEG for at least 20 min on two successive occasions at an interval of 6 hours. Total apnoea and loss of all forms of cerebral activity would then decide in favour of brain death.

Many die-hard ethicists still believe that brain death criteria may not be equated with somatic death. In fact, Japanese do not agree at all with the brain death criteria in their practice; hence the organ availability in Japan is the lowest amongst the economically robust countries.

Circulatory arrest between 75 seconds and 5 minutes is another criteria used in death declaration for organ harvesting. Although it does not mean that auto-cardiac circulation may not begin as it may, but that the permanent loss of vital functions may lead to irreversible damage to brain and other organs. It is believed that permanent cessation of whole brain function at E7 may be accepted as total death [11].

On April 12, 2013, a British tabloid reported that despite the reservations on giving informed consent, there has been a rise of 50 percent post death donations saving thousands of lives in the past few years. So there is something to think about for the rest of the nations, particularly the Muslim world, who believe in charity as a part and parcel of their faith, but are reluctant to practice what they are preach.

Although Muslim ethicists believe in autonomy as practiced in the West, but like it has been discussed elsewhere the actual application of autonomy is dependant on many factors such as religious faith, culture, norms, etc. In the matters of '*Ajal*' the philosophy of autonomy and free will does not apply *per se*. The life does not belong to the human beings. It is on loan by Allah and to Him it must return and the sanctity of life is supreme in the Islamic teachings. So voluntary demand to assist in dying cannot be condoned in Muslim culture nor practiced by Muslim physicians. The whole concept of freedom of choice fails in the matters of taking life as the harm it may bring to the family and the society would be far greater.

In their article, Aramesh and Shadi concluded the arguments as follows:

"Justifying the stance of advocates of euthanasia on the basis of other factors such as economic reasons, consideration of resources, that could otherwise be utilised by other patients."

And death with dignity does not seem plausible because of the criminal nature of mercy killing from the Islamic point of view [12].

"Of course we have to exclude the situation in which the life support equipments are switched off from a brain dead person, aimed to use the organs for saving the life of another person."

The importance of declaration of death is particularly relevant in the matters of organ transplantation. This subject is discussed in an essay on organ transplantation.

To conclude one might say that there are numerous gray zones that certainly require research, i.e. *Ijtehad*. Further in-depth study and analysis must be carried out by those who are knowledgeable and find their way through the maze of ethical dilemmas presented each day by ever growing technology (Further reading: Medicalethics.imamaimedics.com: 2013;vol.2;issue1).

References

1. Yardley L, Dibb B, Osborne G (2003) Factors associated with quality of life in Meneiers disease. Clin Otolaryngol 28:436–441
2. Kirby SE, Yardley L (2008) Understanding psychological distress in Menier's disease. A systemic review. Psychol Health Med 13:257–273

3. Kirby S, Yardley L (2009) Psychological aspects of Meneier's disease and how to manage them. ENT News 18(1): 56–57

4. Hassan HS, Laura G, William Y (1999) Chapter7. In: William Y (ed) Psychobiology of post traumatic discarders- concepts, and therapy. Wiley-Chichester, New York

5. de Pdamal S (1999) Cultural aspects of posttraumatic disorder. In: Yule W (ed) Posttraumatic disorders, concepts and theories. Willey, New York, pp 116–138

6. Charney DS, Deutch AY, Krystal JH, Southwak SM, Davis M (1993) Psycobiologic mechanisms of posttraumatic disorder. Arch Gen Psychiatry 50:294–305

7. Zahedi F et al (2007) End of life issues and Islamic views. Iran J Allergy Asthma Immunol 6(Suppl 5):5–16

8. Al-Qaradwi Y (2007) Quoted from Aramesh and Shadis paper on Eutanasai. Iran J Allergy Asthma Immunol 6(Suppl 5):35–38. http://www.Islamonline.net in Islam on site

9. Khamenei A. Islamic jurisprudence. http://www.basijimed.com/estaftaat/akham-kh/detail.asp quoted by Aramesh and Shadi

10. The sanctity of human life: in Islamic code of ethics. In Islamset.site http://www.islamset.com/ethics/code/cont2.html

11. Rady MY et al (2009) Islam and end of life practices in organ donation for transplantation. HEC Forum 21(2):175–205

12. Ebrahimi AFM (2001) Organ transplantation, euthanasia cloning, animal experimentation; an Islamic view. The Islamic Foundation, Leicester

Further Reading

Medicalethics.imamaimedics.com (2013) 2(1)

Physician's Oaths and Lessons from Masters

10

10.1 Lessons for a Physician in *Dua e Makaram al Akhlaq*

The noble *Quran* is the first and foremost book on ethics. *Nahjal Balagha* is the second document on principles of ethics. Prophet Sa and his most important pupil Ali As are the models of applied ethics and teachers. They are a constant source of referral in all matters of morality and ethics. Imam Syed Sajjad Zainal Abedin is the foremost guide in all matters of prayers and servitude. His *Sahifa Kamilla* is a document that most of Shias read from time to time.

The aim of this chapter is to learn practical tips to become a good human being – and thus a moral physician – from Syed e Sajjad, through *Dua e Makram al Akhlaq*. It is quoted from Imam Jafar Al Sadiq, "As that *Ayma* provides guidance through generic principles and it is for the followers to draw specific conclusions from them." Maulana Kausar Niazi was a great scholar of Islam and Arabic. He once said that the language and diction of the *Quran*, *NahjalBalagha*, and *Saheefa e Kamilla* is identical.

The opening phrases of *Dau e Makaram al Akhlaq* direct us to offer *salawat* upon Mohammed o Aley Mohammad and take our *Emaan* to ultimate heights and to provide strength to *yaqeen* (faith), and make our intention (*niyat*) worthwhile and make our acts reach ultimate heights. He then goes on to advise us that 'may Allah grant my approach free of greed and coercion, and make my faith stronger and remove the deficiencies in my acts.'

In fact, the sum total of this *Dua* is that the Imam has advised us to concentrate on building our character. He has outlined numerous steps to reach perfection. No doubt mankind is imperfect and we are ordinary mortals. We can hardly become perfect; but that is the target at which we are obliged to aim.

The pious Imam has duly highlighted the importance of *Emaan* in his opening remarks, which is exactly what is expected that a Muslim physician to possess: he is expected to have total faith in his Creator and seek His guidance and support in all matters of healthcare. The Muslim physician's oath duly points out that we approach a patient with Bismillah, seeking help from Allah ST. Perfection in achieving total *Emaan* is what all Muslims must strive for – and physicians more so – because we can better understand our fragility and mortality in all states of diseases and sickness. *Emaan* is the pathway towards achieving higher goals in life as physicians. Our responsibility towards our patients, our commitment, sincerity, and honesty and integrity to serve with the only intention of offering the best treatment, avoiding all forms of greed, selfishness dishonesty, theft or torture must be our principles.

Yaqeen is the second aspect that Imam has directed us to observe and excel in. The definition of *Yaqeen* is best given by Imam Ali As in his famous sermon in the *Nahjalbalagha*, which I have discussed elsewhere. Basically, *Yaqeen* implies total belief in an item, subject, topic or entity, so much so that no doubt may remain about it. In fact, there is not much of a difference between *Yaqeen* and *Emaan*. Both ensure our faith in human life, its adversities, the inevitability of death, and so on. For a physician, the wealth of *Yaqeen* is simply the most valuable tool with which to practice medicine. Prophet SA has defined *Yaqeen* as the ultimate stage of *Emaan*, '*Al Yaqeen al Eamaan Kulaahu.*'

Scholars have described three stages of *Yaqeen* in the *Tafseer e Daue Makram Al Akhlaq*. The earliest stage of *Yaqeen* is when one notices evidence of an event and believes in the underlying cause. In other words, the effect may be a source of guidance towards a cause. For instance, one may see a veil of smoke and believe that there must be a fire without actually witnessing the fire. Like the saying goes, "one can't have smoke without a fire."

The present-day model of evidenced-based medicine is a good example of this stage of faith. One may not see any evidence per se, but because of the indicators witnessed by the physician, he or she may presume an underlying cause and proceed further to actually witness evidence such as an abscess or wound or a noticeable pathology. That would be the second stage of *Yaqeen*, that is, one may actually see a fire where the smoke was and ascertain the event. As the saying goes, "seeing is believing."

The final stage of *Yaqeen* is that one may actually feel the fire by touching an ember. That would obviously leave no doubt at all – in medical terms, this stage manifests once the abscess has been noted and aspirated for culture and sensitivity as well as relief for the patient. All three steps in *Yaqeen* are often witnessed in a physician's life.

No physician can succeed in his practical life unless he has confidence in himself, commitment and total faith in his profession, and belief in the ultimate healing power of Allah St.

The third piece of advice the pious Imam gave is the principle of *Niyat*. The importance of *Niyat* or intention in Islam is duly highlighted in all forms of *Ibadaat*. Without a proper *Niyat* even a *namaaz* would not be correct despite the sincerity of offering it. A physician is expected to have clear intention of treating a patient with beneficence in his practical life, without which he may not be able to be fair, honest and earnest, resulting in a possible disaster. Every act in Islam begins with a *Niyat*. For instance, at the beginning of each *salat*, we do a *Niyat*. Likewise, for fasting during the month of Ramadan or *Umrah* or *Hajj*, one must begin an act with a *Niyat*.

In medical practice, the importance of *Niyat* (intention) is obvious to each physician in practice. Every time he sees a patient, he is committing himself to perform his duties honestly, diligently and with the clear intention of benefitting the patient. In fact, physicians all do a *Niyat* when they see a patient on a daily basis. It is customary and perhaps mandatory that a physician should begin his day with the name of Allah ST and seek His blessings at all times, as Allah alone is the healer. The noble *Quran* has mandated that Allah ST is the one who brings malady and He alone can provide a remedy for it.

Scholars have defined the *Niyat* as a connection or an intermediary between Ilm, that is, knowledge, and *amal,* that is, act. It is not essential that one say specific words when making *aniay at.* It is the intention inside one's mind that is more relevant. If the intention is clear and honest, then the moral principles would remain intact, but if a *Niyat* is only spoken or an act such as charity only done to show off, then the purpose of the act would be lost. Such a principle is applicable to physicians in every day life. The whole philosophy of doing good for a patient, if with clear intention and without display of one's authority, may fetch far more rewards in the eyes of Allah than a display of authority and showing off in the care of patient, so often seen among physicians on the subcontinent and surely elsewhere. Somehow a doctor reaching the top of the ladder becomes Pharonic in many such nations and literally begins to act like that Biblical character and the symbol of evil in the noble *Quran.*

Humility and submission to the authority of Allah ST is one of the fundamental attributes that a Muslim physician must possess. *Dua e Makaram Al Akhalaq* duly highlights this fact at many places. In fact, Imam As was known to prostrate himself so often that he used to develop calluses on his forehead. He would submit himself to the authority of Allah ST all the time, praying for forgiveness, mercy, blessings and total obedience. Imam is therefore called Syed ul Sajedin, as he had submitted himself to the total authority of Allah ST. In fact, one of the prayers that he offered and pleaded in the *salat* was begging Allah to grant him his humble daily bread through such means that he may not get entangled in its chase, thus compromising *Ibadat.* Now when we translate this prayer of Imam in the daily life of a Muslim physician, the parables become so significant that one is simply amazed at this supplication. Those of us who have busy medical careers and huge practices, running between surgeries or clinics, to enrich ourselves to raise our standard of living even higher, would appreciate much how often do we compromise our prayers, let alone *Zakat, Umrah* or *Hajj.*

Imam has clearly guided us through this *Dua* when he pleads to Allah ST "let my business not neglect my prayers, and to act ethically and morally so that on the day of judgment I may be able to justify my acts." Physicians cannot ignore this plea because we are answerable to our conscience and conscience is an occult element that in fact is an attribute that only Allah St is aware of. He goes on to plead to Allah, begging Him to grant him independence of any other authority except that of Allah himself. How true in the subcontinent, where sycophancy is an art that is mastered by many physicians attempting to out-do each other to serve the health secretary or a minister. In fact, an old friend called such folks urocrats, as against the bureaucrats. The latter run the country through sycophancy and intrigues, the former carry a sample of the bureaucrat's blood to please him and maintain his own tyrannical position in medical colleges and government hospitals.

Imam goes on to say, "O Allah enable me to serve the mankind with beneficence," which as we know is one of the pillars of medical ethics. He also says that "once I have served the mankind, pray save me from boasting about it." How clearly applicable this is to a physician. It is a distinct difference between an American scientist and a British one, let alone others. The former is usually boastful and arrogant and the later usually modest and down to earth. The discoverer of the World Wide Web, a British scientist Berbers-Lee dedicated it to mankind and did not seek a royalty or rights at all, which could have made him richer than Bill Gates.

Imam AS also pleads to Allah to raise one's stature in public, but simultaneously belittling in one's own eyes; an attribute that we as Physicians fully understand as we desire to excel in our profession, acquire huge stature in the society, both at individual as well as the level of the community, and amongst his peers. But the important message that Imam gives in his supplication is that one may achieve the highest status but simultaneously request to be humble in one's own eyes. What ultimate degree of humility would that bring, and how much respect such a physician would gain, is obvious. There are numerous examples of such people in our profession, who are living models of the plea that Imam has made in this supplication.

At another place, the Imam pleads to Allah St that "when I grow old, provide me with largesse from your treasures and save me from asking anyone else for favours." How very apt for us all irrespective of the profession, as we all have to grow old. As physicians we are blessed with adequate savings and do not normally need economic support from social services, but have we not seen unpleasant examples where either due to a disability or unforeseen happenings, a physician may be exposed to financial constraints; therefore Imam's supplication is pertinent to the physicians as to any other person.

At another place the pious Imam pleads to Allah that if I become well off due to your perpetual blessings, let me not forget that very fact that it is all due to your grace, and pleads Allah not let him become arrogant, ignore His subjects and Ibadat. How very applicable to a physician, who may have made huge sums of money over a long period of time, leading to arrogance, display of wealth, and even turning away from their own kith and kin.

Many physicians do charitable acts, which are quietly performed on regular basis and proverbially speaking even the right hand does not know what the left hand is doing. But many are known to do charitable acts with much pomp and glory, displaying their acts on TV or publishing in the media, more for impressing the peers or the community than act itself. Islam does not endorse an act of charity that is designed to show off. But does it not happen in our daily lives? One can quote several examples where a rich physician may do a fine charitable act, but lose its worthiness by advertising it in the media. Besides, there are many illustrations of rich physicians, who forget their humble beginnings, once they have reached certain status in the society. They may not just ignore their close relatives, even parents, as that may disclose their synthetic current life. I know of one wealthy man in Karachi, who would not acknowledge his real brothers and sisters, even hide his mother, as she was like all mothers, simple and homely. Imam has clearly advised us against such evil acts and begged Allah to save us from such unbecoming attitudes.

In this *Dua*, the pious Imam has given undue importance to 'Amal', i.e. act, action, or skills. Islam grants major importance to Amal along with Ilm. Ali AS said that "knowledge without skills is akin to a bow without an arrow." A physician must have adequate knowledge with matching skills, so as to serve the mankind with beneficence and without malice or non-malificence.

The importance of acquiring *Ilm* and continuing to excel in ones field is duly highlighted by the Imam in the *Dua*, as did his illustrious grandfather Imam Ali AS in his sermons. In fact Ali As gave a statement about knowledge which is simply unmatchable. He said "all pitchers (*Kooza*) fill up when filled, except the pitcher of knowledge, which keeps expanding as filled."

Every physician knows how important it is to learn and keep learning throughout one's life as the knowledge base must continue to extend and grow lest one may become obsolete or outdated in the profession. In fact modern philosophy of medical education is based on this very fundamental principle that 'core knowledge' should be imparted in the medical school, and 'more knowledge' throughout a physicians life. We all know how essential it is; hence all of us are perpetually engaged in continuing medical education, affectionately called CME. But Imam Ali had also warned us that we must not allow ourselves to submerge in the ocean of knowledge, as we may lose the direction in the abyss. Indeed Ali AS categorically advised one to excel in the piece of knowledge that interests one most. How apt an advice for a physician, particularly in today's world of specialisation, and super specialisation. Remember the old saying, "Jack of all trades, master of none."

Justice is the most important pillar of medical ethics. A median drawn between excess and deprivation (*Ifrat* o *Tafreet*) is one of the definitions given by the exponents of *Dua e Makaram Akahlaq*. Justice is in fact the fundamental brick of ethics *per se*. Scholars have identified four components of morality. They are: wisdom, piety, courage and justice.

Although all elements mentioned above are important in their own right, justice is the bridge that joins all these elements into a formidable entity, i.e. morality. Any deviation from the centre point or fulcrum would stabilize the equation, and may lead to its demolition.

For instance wisdom is a virtue, and a hugely meritorious attribute, but only if it remains within specific proportions compatible with the principle of justice. Its excess will turn a person into a cunning and devilish person. Likewise if the wisdom falls from altar the person may become a buffoon. A physician should be wise and not cunning; should employ his *aql* and not be a clown.

Similarly, one of the attributes that a physician must possess is that of leadership and entrepreneurship. In fact the GMC in its periodical publications keeps reminding its members that apart from observing the principles of Good Medical Practice, a physician should also possess leadership qualities. And the basic ingredients of leadership are the attributes of courage and bravado. Imam has given due importance to courage as an important attribute for his followers to own and possess. But we must also remember that it's essential that even courage must be in balance with other attributes If there is excess in the attribute of courage, it may become arrogance or terror for the subordinates, and if it is less than optimum it may be labelled as cowardice. Many of us have to work in small rural areas, where a physician is second only to the village *parson* or a *mullah* in terms of honour and dignity. He may thus have to face many a challenge, for which courage, foresight, clarity of mind and power of decision making are necessary attributes.

It is therefore expected of a physician to justify his status by showing leadership qualities with courage and bravado, so often needed in a remote region, where he may be the only authority to make life and death decision. The village folk might turn towards him for their domestic and community issue for a wise counsel, and he may be obliged to take decisions that one party may not like. It may therefore sow the seeds of animosity.

Justice maintains a balance between all the attributes mentioned above. Any deviation from justice may lead to loss of a physician's integrity and authenticity. If justice is compromised and matters taken to an extreme, it may make one arrogant, egotistic, and outright *zalim*, which is an exact opposite of *Adl*. A physician would be totally unethical if he does not observe justice in day to day life, as defined by *ulema* in the light of imam Sajjad's supplications.

The fundamental rule of ethics is to follow the straight path as commanded by Allah ST in "*haza siraty mustaqeem fantabulous, wala tatabau al sabeel catafalque bikum Uni sabeelahu.*" Translated as "verily this is my straight path, that you must follow, and avoid other paths, that may misguide you from the the rightful ways" (*Tafseer e Dua e Makaram alAkhlaq*).

Finally one must discuss the attribute of piety that Imam has emphasised above all attributes. *Taqwa* is one basic ingredient that Quran, Sunnah and the teachings of Ahlalbeit have repeatedly commanded all Muslims to observe and practice. Taqwa is the end product of the total submission to the authority of Allah ST. In the *Tafseere Dua makarama alAkhlaq*, the ulema have mentioned numerous *ayats* in this regard such as "that if you are patient and observe piety (*taqwa*) then (know) that these are courageous acts."

Another *ayat* translates as that "if you observe pertinence and piety, others' cunning will not bring any harm unto you." And that "Allah is with all those who are fearful of Him." Taqwa saves you from strictures, on *Rizq*, and expands the largesse."

Taqwa is also a means of improvement in one's actions. "Be fearful of Allah ST and speak truth, so that Allah may rectify your acts." Then there is the famous mandate in Quran "*In- Allah youhibul mutaqeen.*" It means that Allah counts the pious people amongst His friends.

Piety is also a means to an end of approval of supplications by Allah ST. It is also means to an end of respite from hell, and abyss.

At another place Allah has commanded that "verily one who is pious amongst you is closest to Allah ST."

Is there a physician who would not attempt to follow these commandments given by Allah ST. Perhaps not all of us are good enough to reach the highest perimeters, but at least we must endeavour and do our utmost to archive *Taqwq*, thus serving our patients with the finest attribute dictated by Allah ST.

Three fundamentals of religion described by Muslim scholars are, *Din* ie Islam, *Emaan* and *Ehsaan*. Islam is the founding brick followed by *Emaan*. It means that every Muslim is not necessarily a momin. It has been mandated in *Quran* per se. So *Emaan* is a higher grade than Islam, but *Ehsaan* is the highest of all three grades.

Imam As in this *Dua* speaks of *Ehsaan* many times. *Ehsaan* is a higher attribute than *Adl*. It comprises of all those elements mentioned above under the caption of Adl plus sacrifice, altruism and going an extra mile to serve the mankind. In *Surat al Rahman*, Allah has commanded as "*Hal Jaza al Ehsaan illaehsaan*", i.e. can there be another substitute for *Ehsaan* than *Ehsaan*, simply highlighting the importance of *Ehsaan* in Islam. Aqeel Gahravi is an *Alim e din*, and a philosopher. He gave a fine example of *Ehsaan* in a lecture that when the *Ahal Beit* fasted on three successive days and gave away their daily bread to a beggar, who in fact was an angel sent by Allah to check the attribute of patience, tolerance and altruism. Allah loved their act of *Ehsaan* and bestowed them with the gift of *ayat* e "*Al-Dahr*."

Likewise when a beggar asked for help, while all *sahaba* were praying behind the Prophet SA, no one but Ali extended his hand out to the beggar to withdraw the ring from his finger. This act was an act above and beyond the call of duty, and was a typical example of *Ehsaan*.

A physician must dispense justice in all his professional duties, but to go an extra mile to help the destitute and deserving few with free service or providing medication, distributing artificial appliances like eye glasses or hearing aids, or limb prosthesis all in the name of service to mankind is certainly an act of *Ehsaan*, that Allah ST would simply love and grant the physician further success in his profession.

Imam AS has also advised us to ignore the shortcomings of peers or friends, as it may otherwise lead to bitterness and eventually breakdown of the relationship. It is so important for a physician that he observe this rule in daily life as no one is perfect. Causing emotional trauma to the one who may be less capable than yourself by highlighting his deficiency will bring harm to friendship. In fact it is often seen in physicians who are legends in their fields that they tend to do '*chashm poshi*' of a less capable person and guide him in private to rectify the deficiencies, rather than ridicule him in public and without causing unpleasantness to the person concerned. In fact, many professors or even higher academicians like Nobel Laureates are so simple and down to earth that one almost naturally recalls the saying of Imam Ali As, when your meet such a one that "be like a tree that bears fruit and provides shade, to serve the mankind and not a tree that may not bear even a leaf, let alone fruit or provide shade." As a matter of fact such physicians and scientists, or any person of academic excellence is usually humble, modest and down to earth. The one who is arrogant or haughty and proud of status in academics or profession is a shallow human being, who may suffer humiliation at some stage in his life due to his condemnable traits.

A physician who highlights the deficiencies of a colleague overtly or covertly, or indeed keeps a record of such deficiencies for exploitation and does not rectify him, is obviously an evil person. Surely we have all come across such colleagues in our lives. They are petty, myopic, cunning and basically bad guys. One must avoid such evil traits in order to be a good doctor. Imam has clearly warned against such traits and begged Allah in this *Dua* to grant him the habit of ignoring others' weaknesses and shortcomings. Allah ST ignores and hides our thousand shortcomings, and we must do the same with our peers, colleagues and human beings in general. It is an absolutely vital attribute for a physician to possess.

Imam gives special instructions on behaving respectfully and honourably with friends, colleagues, peers and juniors in all our matters. Inter-professional relationship is a huge subject in medical ethics. It is essential that all physicians behave with mutual respect and dignity amongst themselves. Duly recognise the calibre and expertise of a professional colleague, giving due importance to their opinion about a certain situation and keeping it confidential, with trust and faith. These are some on the basic components of inter-professional ethics.

All philosophers from Hippocrates and Galen to the President of the Royal Colleges or the Chair of the GMC have advised indeed commanded a physician to possess mild and pleasant manners, and soft demeanour; be polite, kind and generous.

In *Dua e Makaram Al Akhlaq*, Imam AS has begged Allah to grant him soft and mild manners, politeness and pleasantness in dealings with human beings. Such an attitude brings about joy in life of both parties, namely yourself as well as those with whom you are dealing. In fact, Imam Ali As said in one of the sermons that a man must treat his family like a bird treats its wings, with love and care, as these wings enable it to fly high in the skies.

The patients' family is like his own family to a physician, who must be treated with love, affection, devotion, empathy and the utmost care he can provide.

Imam gives huge importance to 'truth' in one's life. No attribute is as valuable as truth telling in the life of a physician. Truth is the building block of an individual's character. It gives depth to an individual and grants him depth in personality, and helps him serve mankind more intensely and honestly. Truth is a major subject in a physician's life and has been discussed elsewhere in this book.

In *Dua e makarame Akhlaq, all* those attributes are discussed by Imam As that we as physicians are obliged to possess. The lessons that I have learned are as follows.

1. A physician must possess a sound character. Humanity is the first and foremost attribute for a physician.
2. He should be humble, modest, and down to earth.
3. Must know his limitations and ignore others' shortcomings and weaknesses. He should not exploit them to ridicule or belittle them.

4. Must possess due knowledge and matching skills, as Ilm o Amal are true companions. Excess of knowledge may make one a nerd and the excess of skills without matching knowledge may convert him into a technician.

5. A physician must have pleasant demeanour, soft voice, smiling face and humility in his walk, talk and action (remember what the wise man *Luqman* instructed his son, in *Quran*!).

6. He must give the best care to his patient treating him like his own family, listening to him patiently without undue hurry. Respect his autonomy and human rights. Give due regards to confidentiality, anonymity and choice of treatment; without greed, malice or avarice. Autonomy, birr or beneficence and non-malificence are the fundamental bricks of medical ethics.

7. Must not abuse his authority, and should always use his skills for the welfare avoiding harm or damage to his patients.

8. Earn an honest living, so that he is comfortable and not be at the mercy of others. He should ask Allah only for His sustenance, so that he may not forget the *salat*, and other *Ibadaat*, in the pursuit of his livelihood.

9. He must avoid bickering, unpleasantness, backbiting or sycophancy. He must have full faith in his capability and total *Yaqeen* in the fact that Allah has all the power to heal his patients. The physician's only task is to leave no stone unturned to save a life.

10. Death is the ultimate truth. All creatures have to meet it. Until that happens, a physician must not refrain from doing his best to take care of the sick, and the dying.

11. A physician must maintain a balance between various attributes that he may possess, such as knowledge and skills, courage and adventurism, empathy and antipathy, earning an honest livelihood and accumulation of wealth, etc.

12. A physician must remain actively engaged in the pursuit of knowledge all his life. Be a leader and an entrepreneur.

The characteristics of a Muslim physician have been described in a nice article (*Islamic medical educational resources by Mohammadiyah Yogjakarata and forum Kedoketran Islam Indonesia 24–25 Aug. 2007*), which is worth a read.

It says that the Prophet gave general rules of morality which can guide the medical personnel in their daily lives. Some of these principles are as follows.

- Have total faith in Allah ST (*Yaqeen*)
- All work is duly recognised according to the intention (Niyat).
- Doubtful actions must be avoided, i.e. have faith (Yaqeen)
- Leaving alone what is not yours (Ali AS was once asked as to where did he find peace. He replied "in keeping away from others' affairs").
- La dharar wa la dharrara (beneficence and non-malificence).
- Sincerity in advice.
- Renouncing desire for material gains.
- Conscience must decide between right and wrong (justice is the tangible form of conscience).
- The right acts please the heart and evil bring distress.
- Anger and rage must be avoided (Ali said, "anger commences with stupidity and culminates in remorse").

- Guard your tongue (politeness, softness, and cultured conversation). Ali AS said, "a man is hidden behind his tongue."
- Do good and avoid evil.
- Restraint and modesty should be your treasures.
- Courage to speak truth in the face of adversities.
- Wisdom or *Hickma*.
- Patience and *sabr.*
- Humility or *tawadhu'u.*
- Self restraint.
- Modesty or *haya.*
- Simplicity and avoidance of worldly breeds.
- Moderation and balance in dealing with people, in your expenditure and in decision making.
- Enterpreunial attitude in life.

The author has quoted nearly a hundred references from *Quran* and *Sunnah*, quoting, *Sahi Bukhari, MuslimTirmidhi, Nisai, Ibne M*ajah and many other extremely authentic Muslim sources.

To conclude, one may say that the message of Allah St was delivered to us by Prophet Mohammad Sa and his Ahla Beit. Imam Syed e Sajjad has followed exactly those guidelines in *Dua e Makaram Al Akahlaq* that his grandfather Prophet SA gave us in his *ahadith.*

Now it is up to a Muslim physician to follow them and be rewarded or ignore them and be cursed. The choice is simple!

10.2   Views of Some Famous *Muslim* Scholars on Medical Ethics

Ali iIbne Rabban al Tabri (838–870 AD) advised the physician as follows: "people grant high rank to a kind and generous. Therefore it is essential for a physician to be kind, generous, contented and possess a sound character. A physician should be more kind to his patients than even to his own family, and put them ahead of his personal self. He should act more than talk. Neither should he be proud, nor greedy for money. He should only prescribe tried and tested medicine."

Another great Muslim physician, al-Razi (865–925 AD) a contemporary of Al-Kindi. Who lived in Koofa wrote as follows: "a physician should have total faith in the art of healing. He must never try to achieve greatness, dignity and honour by humiliating others. He should always be on beck and call of poor patients (as well as the well to do), as it is a noble act. Although a physician may not be sure of the patient's recovery, he must reassure him of swift recovery, as it will assist the patient's natural healing power nature (Vis medicatrix nature)."

And finally most noted of them all, Ali Husayn-Abdullah ibne Sina (Avicenna), who observed the principles of ethics diligently in his profession, respecting and protecting human life as commanded in Quran. He experimented and conducted research perusing knowledge and skill all his life. He wrote, "A physician should be true to his faith in Islam and its precepts. He should be honest to himself, his profession and his patient." Historians have mentioned that Avicenna kept himself in line with the Muslim *Mutakallimun* (Dialecticians). To him, Allah was the ultimate healer, a tract that he duly demonstrated in his speeches and writings. He said that a physician should be soft spoken and polite; these being the essential attributes for a physician.

An ideal physician, in the light of the Muslim philosophers' teachings and practice should treat his patients honestly and selflessly, without greed or avarice, fearing Allah and keeping Him in his thoughts all the time while employing his knowledge and skills to treat his patients. He should honour the privacy and confidentiality, and treat him with sympathy, and without any motivation for greed or exploitation. He must know his limitations endeavour to excel, remove the deficiencies and remain constantly in touch with his peers, whom he must respect.

10.3 Characters of a Muslim Physician

"The physician should be amongst those who believe in God, fulfil His rights, are aware of His greatness, are obedient to His orders, who refrain from His Prohibitions, and who serve Him in secret and in public.

The physician should be endowed with wisdom and practise graceful admonition. He should be cheering not dispiriting, smiling and not frowning, loving and not hateful, tolerant and not edgy. He should never succumb to a grudge or fall short of clemency. He should be an instrument of God's justice, forgiveness and not punishment, coverage and not exposure.

He should be so tranquil as never to be rash even when he is right. Chaste of words even when joking … tame of voice and not noisy or loud, neat and trim and not shabby or unkempt … conducive of trust and inspiring of respect … well-mannered in his dealings with the poor or rich, modest or great. In perfect control of his composure … and never compromise his dignity, however modest and forbearing.

The physician should firmly know that 'life' is God's … awarded only by Him … and that 'Death' is the conclusion of one life and the beginning of another. Death is a solid truth … and it is the end of all but God. In his profession the physician is a soldier for 'Life' only … defending and preserving it as best as it can be, to the best of his ability.

The physician should offer a good example by caring for his own health. It is not befitting for him that his 'do's' and 'don'ts' are not observed by himself. He should not turn his back on the lessons of medical progress, because he will never convince his patients unless they see the evidence of his own conviction … God addresses us in the '*Quran*' by saying, 'and make not your own hands throw you into destruction.' The Prophet says, 'Your body has a right on you' … and the known dictum is 'no harm or harming in Islam.'

The physician is truthful whenever he speaks, writes or gives testimony. He should be invincible to the dictates of creed, greed, friendship or authority pressurising him to make a statement or testimony that he knows is false. Testimony is a grave responsibility in Islam. The Prophet once asked his companions, "Shall I tell you about the gravest sins?" When they said "yes," he said "claiming partners with

God, being undutiful to one's parents … " and after a short pause he repeatedly said "and indeed the giving of false talk or false testimony."

The physician should be in possession of a threshold-knowledge of jurisprudence, worship and essentials of *Fiqh* to enable him to give counsel to patients seeking his guidance about health and bodily conditions with a bearing on the rites of worship. Men and women are subject to symptoms, ailments or biological situations. For instance, women during pregnancy may wish to know the religious ruling pertaining to prayer, fasting, pilgrimage, family planning, etc. Although 'necessity overrides prohibition' the Muslim physician-nevertheless-should spare no effort in avoiding the recourse to medicines or therapy or surgery, or medical or behavioural dictates that are prohibited by Islam.

The role of a physician is that of a catalyst through whom God, the Creator, works to preserve life and health. He is merely an instrument of God in alleviating people's illnesses. For being so designated, the physician should be grateful and forever seek God's help. He should be modest, free from arrogance and pride and never boast or hint at self-glorification through speech, writing or direct or subtle advertisement.

The physician should strive to keep abreast of scientific progress and innovation. His zeal or complacency and knowledge or ignorance, directly bear on the health and well-being of his patients. Responsibility for others should limit his freedom to expend his time. As the poor and needy have a recognized right in the money of the capable, so the patients own a share of the doctor's time spent in study and in following the progress of medicine.

The physician should also know that the pursuit of knowledge has a double indication in Islam. Apart from the applied therapeutic aspect, pursuit of knowledge is in itself worship according to the Qur'anic guidance, "And say … My Lord … advance me in knowledge" and "Among His worshippers … the learned fear Him most", and "God will raise the ranks of those of you who believed and those who have been given knowledge."

(*Source: Islamic Code of Medical Ethics Kuwait Document, International Organization of Islamic Medicine 1981*).

10.4 A Muslim Physician's Oath

"In the name of Allah, Most Gracious, Most Merciful".

Praise be to Allah, the Sustainer of His Creation, the All-knowing.

Glory be to Him, the Eternal, the All-Pervading.

O Allah, Thou art the only Healer, I serve none but Thee, and, as the instrument of Thy Will, I commit myself to Thee.

I render this Oath in Thy Holy Name and I undertake:

To be the instrument of Thy Will and Mercy, and, in all humbleness, to exercise justice, love and compassion for all Thy Creation;

To extend my hand of service to one and all, to the rich and to the poor, to friend and foe alike, regardless of race, religion or colour;

To hold human life as precious and sacred, and to protect and honour it at all times and under all circumstances in accordance with Thy Law;

To do my utmost to alleviate pain and misery, and to comfort and counsel human beings in sickness and in anxiety; to respect the confidence and guard the secrets of all my patients;

To maintain the dignity of health care, and to honour the teachers, students, and members of my profession;

To strive in the pursuit of knowledge in Thy name for the benefit of mankind, and to uphold human honour and dignity;

To acquire the courage to admit my mistakes, mend my ways and to forgive the wrongs of others;

To be ever-conscious of my duty to Allah and His Messenger (S.A.W.), and to follow the precepts of Islam in private and in public.

And finally the original, traditional and ageless Hippocratic oath.

The Oath

By Hippocrates

(Written 400 B.C.E)

I swear by Apollo the physician, and Aesculapius, and Health, and All-heal, and all the gods and goddesses, that, according to my ability and judgment, I will keep this Oath and this stipulation – to reckon him who taught me this Art equally dear to me as my parents, to share my substance with him, and relieve his necessities if required; to look upon his offspring in the same footing as my own brothers, and to teach them this art, if they shall wish to learn it, without fee or stipulation; and that by precept, lecture, and every other mode of instruction, I will impart a knowledge of the Art to my own sons, and those of my teachers, and to disciples bound by a stipulation and oath according to the law of medicine, but to none others. I will follow that system of regimen which, according to my ability and judgment, I consider for the benefit of my patients, and abstain from whatever is deleterious and mischievous. I will give no deadly medicine to any one if asked, nor suggest any such counsel; and in like manner I will not give to a woman a pessary to produce abortion. With purity and with holiness I will pass my life and practice my Art. I will not cut persons labouring under the stone, but will leave this to be done by men who are practitioners of this work. Into whatever houses I enter, I will go into them for the benefit of the sick, and will abstain from every voluntary act of mischief and corruption; and, further from the seduction of females or males, of freemen and slaves. Whatever, in connection with my professional practice or not, in connection with it, I see or hear, in the life of men, which ought not to be spoken of abroad, I will not divulge, as reckoning that all such should be kept secret. While I continue to keep this Oath unviolated, may it be granted to me to enjoy life and the practice of the art, respected by all men, in all times! But should I trespass and violate this oath, may the reverse be my lot!

Further Reading

Al Tibb al Islami Hakim Mohd Said Hamdard Foundation Pakistan (1972) pp 73–77

Avicenna and medical ethics. A paper presented by Hakim Mohd. Said at the international congress of history of science, Bucharest, 26 Aug 3 Sept (1981)

Medical ethics in Islam Hakim Abdul Hameed, Hamdard Medicus (1978) xxi:7–12. pp 3–16

Zaidi S (2001) Ethics of medicine. Pakistan J Otolaryngol 17(Suppl):1. Karachi, Pakistan

Index